Conducting Organic Materials and Devices

SEMICONDUCTORS
AND SEMIMETALS
Volume 81

Semiconductors and Semimetals

A Treatise

Edited by R.K. Willardson
    CONSULTING PHYSICIST
    12722 EAST 23RD AVENUE
    SPOKANE, WA 99216-0327
    USA

Eicke R. Weber
DEPARTMENT OF MATERIALS
SCIENCE AND MINERAL
ENGINEERING
UNIVERSITY OF CALIFORNIA
AT BERKELEY
BERKELEY, CA 94720
USA

Conducting Organic Materials and Devices

SEMICONDUCTORS
AND SEMIMETALS

Volume 81

SURESH C. JAIN
New Dehli, India

MAGNUS WILLANDER
Göteborg and Linköping, Sweden

VIKRAM KUMAR
New Dehli, India

AMSTERDAM • BOSTON • HEIDELBERG • LONDON • NEW YORK • OXFORD
PARIS • SAN DIEGO • SAN FRANCISCO • SINGAPORE • SYDNEY • TOKYO
Academic Press is an imprint of Elsevier

Academic Press is an imprint of Elsevier
84 Theobald's Road, London WC1X 8RR, UK
Radarweg 29, PO Box 211, 1000 AE Amsterdam, The Netherlands
Linacre House, Jordan Hill, Oxford OX2 8DP, UK
30 Corporate Drive, Suite 400, Burlington, MA 01803, USA
525 B Street, Suite 1900, San Diego, CA 92101-4495, USA

First edition 2007

Copyright ©2007 Elsevier Inc. All rights reserved

No part of this publication may be reproduced, stored in a retrieval system or transmitted in any form or by any means electronic, mechanical, photocopying, recording or otherwise without the prior written permission of the publisher

Permissions may be sought directly from Elsevier's Science & Technology Rights Department in Oxford, UK: phone (+44) (0) 1865 843830; fax (+44) (0) 1865 853333; email: permissions@elsevier.com. Alternatively you can submit your request online by visiting the Elsevier web site at http://elsevier.com/locate/permissions, and selecting *Obtaining permission to use Elsevier material*

Notice
No responsibility is assumed by the publisher for any injury and/or damage to persons or property as a matter of products liability, negligence or otherwise, or from any use or operation of any methods, products, instructions or ideas contained in the material herein. Because of rapid advances in the medical sciences, in particular, independent verification of diagnoses and drug dosages should be made

ISBN: 978-0-12-752190-9

ISSN: 0080-8784

For information on all Academic Press publications
visit our website at books.elsevier.com

Printed and bound in USA

07 08 09 10 11   10 9 8 7 6 5 4 3 2 1

**Working together to grow
libraries in developing countries**

www.elsevier.com | www.bookaid.org | www.sabre.org

ELSEVIER   BOOK AID International   Sabre Foundation

Contents

PREFACE	ix
Chapter 1 Introduction	1
1.1. ADVANTAGES OF CONDUCTING POLYMERS	1
1.2. EARLY ATTEMPTS FOR APPLICATIONS	2
1.3. GROWTH AND PROPERTIES	2
1.4. ACTIVE DEVICES	4
Chapter 2 Polyacetylene	7
2.1. STRUCTURE, GROWTH, AND PROPERTIES	7
2.1.1. *Structure*	7
2.1.2. *Growth and Doping of Polyacetylene*	7
2.2. BAND STRUCTURE OF t-PA	9
2.3. THE SOLITONS AND THE POLARONS	12
2.3.1. *The Solitons*	12
2.3.2. *The Polarons*	13
2.4. TRANSPORT PROPERTIES	16
2.4.1. *Mobility in Selected Polymers*	16
2.4.2. *Conductivity and Susceptibility*	20
Chapter 3 Optical and Transport Properties	23
3.1. EFFECT OF ELECTRIC FIELD ON PHOTOLUMINESCENCE (PL)	23
3.2. THE DIELECTRIC CONSTANT	27
3.3. SPACE CHARGE LIMITED CURRENTS	28
3.3.1. *Early Work of Mott. Poisson and Continuity Equations in a Trap-Free Insulator*	28
3.3.2. *Effect of Background Doping*	31
3.4. POLYMERS: THE SOLIDS WITH TRAPS	32
3.4.1. *Poisson Equation with Trapped Charges*	32
3.4.2. *Single Level Traps*	32
3.4.3. *Gaussianly Distributed Traps*	34
3.5. EXPONENTIAL TRAPS	35
3.5.1. *Calculation of $J(V)$*	35
3.6. RELAXATION OF THE APPROXIMATION $p_t \gg p$	35
3.6.1. *J–V Curves When $p_t \gg p$*	35

3.6.2. *Trap-Filled Limit*	36
3.7. EFFECT OF FINITE (NON-ZERO) SCHOTTKY BARRIER	38
3.7.1. *Importance of Finite Barriers*	38
3.7.2. *Theory*	39
3.7.3. *Results and Discussion*	40
3.7.4. *Comparison with Experiment*	42
3.8. COMBINED EFFECT	44
3.9. TEMPERATURE EFFECTS	44
3.9.1. *Temperature Effects in PPV-Based Polymers*	44
3.9.2. *Recent Work*	48
3.9.3. *Temperature Effects in MEH-PPV. Recent Work*	50
3.9.4. *Temperature Effects in Alq$_3$*	51
3.10. MOBILITY MODEL OF CHARGE TRANSPORT	55
3.11. UNIFIED MODEL	55
3.11.1. *Shallow Gaussian and Single Level Traps*	55
3.11.2. *Unified Model with Exponentially Distributed Traps*	57
3.12. HIGH FIELD OR POOLE–FRENKEL EFFECT	58
3.12.1. *J–V Characteristics*	58
3.12.2. *Calculations and Comparison with Experiments*	60
3.13. MOBILITY OF CHARGE CARRIERS	62
3.13.1. *Bulk Materials*	62
3.13.2. *Mobility in Blends*	63
3.14. IMPORTANT FORMULAS	64
3.15. SUMMARY OF THIS CHAPTER	64

Chapter 4 Light Emitting Diodes and Lasers   67

4.1. EARLY WORK	67
4.2. BLUE, GREEN AND WHITE EMISSION	70
4.2.1. *Blue and Green LEDs*	70
4.2.2. *White Light Emission from Organic LEDs*	74
4.3. COMPARISON WITH OTHER LEDS	80
4.4. ORGANIC SOLID STATE LASERS	81
4.4.1. *Photopumped Lasers*	81
4.4.2. *Spectral Narrowing*	82
4.4.3. *Blue Lasers*	83
4.5. QUANTUM EFFICIENCY AND DEGRADATION	83
4.6. STABILITY	86
4.6.1. *Degradation of the Polymer*	86
4.6.2. *The Cathode and the Black Spots*	87
4.6.3. *Degradation of the Anode*	90
4.7. SOLUBLE NEW 5-COORDINATED AL-COMPLEXES	91
4.8. SUMMARY AND CONCLUSIONS	94

Chapter 5 Solar Cells   95

5.1. INTRODUCTION	95
5.2. SOLAR CELLS	95
5.2.1. *Single and Bilayer Solar Cells*	95

	5.2.2. *Interpenetrating Network of Donor–Acceptor Organics. Bulk Heterojunction Solar Cells*	97
5.3.	SOURCE OF V_{OC} IN BHSCS	104
	5.3.1. *Effect of Acceptor Strength [133]*	104
	5.3.2. *More Recent Work [134,135]*	105
5.4.	OPTIMUM PCBM CONCENTRATION	107
	5.4.1. *Superposition Principle*	108
5.5.	MODELING THE OUTPUT CHARACTERISTICS	112
	5.5.1. *The Output Currents*	112
	5.5.2. *The Model*	115
5.6.	COMPARISON WITH OTHER SOLAR CELLS	117
	5.6.1. *Amorphous Si Solar Cells*	117
	5.6.2. *Polycrystalline Si Solar Cells*	119
5.7.	SUMMARY AND CONCLUSIONS	122

Chapter 6 Transistors   123

6.1.	IMPORTANCE OF ORGANIC TFTS	123
6.2.	EARLY WORK	124
6.3.	EFFECT OF TRAPS	126
6.4.	HIGH FIELD EFFECTS	127
6.5.	TRANSPORT IN POLYCRYSTALLINE ORGANICS	130
	6.5.1. *Effect of Grain Boundaries*	130
6.6.	PENTACENE TFTS	131
6.7.	CONTACTS	138
6.8.	ORGANIC PHOTOTRANSISTOR	140
6.9.	ORGANIC DIELECTRICS	143

BIBLIOGRAPHY	147
INDEX	157
CONTENTS OF VOLUMES IN THIS SERIES	167

Preface

Conducting polymers were discovered in the 1970s in Japan. Since this discovery, there has been a steady flow of new ideas, new understanding, new conducing polymer (organics) structures and devices with enhanced performance. Several breakthroughs have been made in the design and fabrication technology of the organic devices. Almost all properties, mechanical, electrical, and optical, are important in organics. Great advances have been made in the understanding of the transport of charge carriers in the conducting organics. High performance LEDs, lasers, solar cells and transistors have also been developed. The performance of organic transistors has become comparable to that of the amorphous Si transistors while the performance and stability of the organic LEDs have become so good that they have already reached the market place. The reported efficiency of the bulk heterojunction organic solar cells is already more than 5%. This book describes the recent advances in these organic materials and devices.

The number of papers which have recently been published on the organic materials and devices has become very large. We have included some 200 papers in the bibliography, which are most relevant for the coherent description of recent developments. To make the bibliography more useful, titles of the most important papers have been included, and the subject matter is treated at an appropriate level for students as well as senior researchers interested in the design and modelling of the organic devices.

The authors have benefited from interaction and collaboration with such a large number of colleagues that it is difficult to mention them all individually. Our discussions with Prof. A.M. Stoneham of University College, London, Prof. A. Atkinson of Imperial College, London, and Dr. M.N. Kamalasanan of National Physical Laboratory, New Delhi, have been very useful. One of us (S.C.J.) has derived considerable benefit from his stay at IMEC and also as a visiting professor of KU Leuven. Mrs. Anubha Jain helped in the organization of the material in this book and made useful suggestions. We are indebted to Mr. Pankaj Kumar for many useful discussions. Mr. Kumar read the whole manuscript and made many useful suggestions.

We wish to extend special thanks to Mr. David G. Sleeman, of Elsevier UK, for his personal support for this book. He always showed the utmost consideration to us. It was due to the skill and efforts of Mr. Sleeman that the book could be completed in time. He deserves our sincere thanks.

Finally, we must thank sincerely our children and our wives who provided us unfailing support and help during the preparation of this book.

S.C. Jain
Magnus Willander
Vikram Kumar
November 30, 2006

CHAPTER 1
INTRODUCTION

1.1. Advantages of Conducting Polymers

Plastics created a lot of excitement in the 1940s and 1950s. These plastics were insulators. Until the early 1970s the idea that plastics could conduct electricity would have appeared absurd. The first conducting polymer was synthesized in the 1970s. In the early 1970s a graduate student in Shirakawa's laboratory in Tokyo was trying to make polyacetylene from the acetylene gas. Instead of the polyacetylene (which is known to be a dark powder), the student produced a lustrous metallic looking film. The film looked like an aluminum film but stretched like a thin plastic sheet. Accidentally the student had added 1000 times more catalyst than the amount required to produce polyacetylene. Subsequently Shirakawa collaborating with MacDiarmid and Heeger of the University of Pennsylvania could increase the conductivity of the polyacetylene films a billion times by doping it with iodine. The doped films looked like golden metallic sheets. Later more than a dozen organic polymers could be made conducting by appropriate doping. Now plastics with conductivity comparable to that of copper can be easily fabricated. Early work on conducting plastics is described in [1–8].

The semiconducting conjugated polymers can be used as the active layer in LEDs, field effect transistors, solar cells, photodiodes, electrochemical cells and memory devices. They have proved to be of great importance as an active medium in lasers. These devices are being pushed toward commercialization because they can be fabricated by inexpensive techniques, such as spin coating, ink-jet printing, low temperature fiber drawing and screen-printing on the flexible substrates. This leads to a real advantage over the expensive and sophisticated technology used with inorganic materials in the semiconductor industry. The glass and flexible plastic foil make these devices particularly interesting because of the advantages they offer in terms of flexibility, low power, low weight, and low cost. In view of the above advantages, conducting plastics have emerged as a new class of electronic materials. It is possible that by the year 2010, silicon might hit the wall and the conducting polymers may become the major players in the field of semiconductor devices.

The polymers consist of chains, each chain contains C–H or related groups bound together by strong sigma bonds, which provide strength and integrity to the polymer. Inter-chain coupling is small. Therefore the materials are quasi-one dimensional (quasi-1D). The structure allows the dopant atoms or molecules to go in space in-between the chains.

1.2. Early Attempts for Applications

Early work on the conducting polymers has been discussed in the reviews [2,9,10]. Early applications of polymers as processible conductors are also given in these publications. There were attempts to commercialize the conducting polymer products. Two Japanese companies manufactured rechargeable button-cell batteries. The batteries used polyaniline and lithium electrodes. These batteries had a longer lifetime than the nickel–cadmium and lead–acid batteries. However the venture was not successful. A textile manufacturer Milliken and Company in the USA developed a fabric known as Contex. Contex had conducting polymer polypyrrole interwoven with other common synthetics. This fabric is excellent for camouflage because it fools the radar by making it appear that the signal is going through empty space. Contex was approved for use but for a variety of reasons the material was not successful. The annual damage to electronic equipment by electrostatic charges is estimated to be more than $15 million. At present the protective packaging relies on ionic salts or resins filled with metals or carbon. The conductivity of ionic salts is low and it is unstable. A metal is expensive and heavy. Carbon bits cast off during shipment and can cause contamination. IBM is developing a polyaniline solution, known as PanAquas. If conductivity of PanAquas could be increased, it could replace the lead based solder, which is hazardous. Polymers are good candidates for electromagnetic shielding also. Allied Signals developed a product named Versacon, which was similar in performance to the IBM PanAquas. Several companies incorporated it in paints and coatings. However the volume of sales continued to be too small and Allied had to stop its production. As compared to Versacon, which is in a powder form, PanAquas is a solution and is transparent. Epstein has a patent on a technique to join two pieces of plastics using conducting polymer polyaniline. The pieces to be joined are sprinkled with polyaniline and irradiated with microwaves. Polyaniline absorbs the energy from microwaves, melts and fuses the two pieces together. MacDiarmid and collaborators have made polymer electromechanical mechanisms. The polymers undergo large changes in dimensions with small electric currents. Potentially several microactuators coupled together could be used as artificial muscles.

1.3. Growth and Properties

Monolayer control of thickness of organic films has been obtained using the Langmuir–Blodgett film deposition technique as well as layers grown by self-assembled monolayer from solutions [11]. However the structure of the films grown by these methods cannot be controlled accurately. Since its emergence in the mid-1980s, Organic Molecular Beam Epitaxy (OMBE) has become an important technique for deposition of organic films with monolayer thickness control in atomically clean environments. Extreme chemical and structural control of the films is also obtained. Forrest [11] has written an extensive review on the structure and properties of the films grown by this method.

Typical conductivity values of several polymers are shown in Fig. 1.1. The conductivity of polymers is compared with the conductivity of other solids in Fig. 1.2. It is seen that the maximum conductivity is quite high and close to that of good metals.

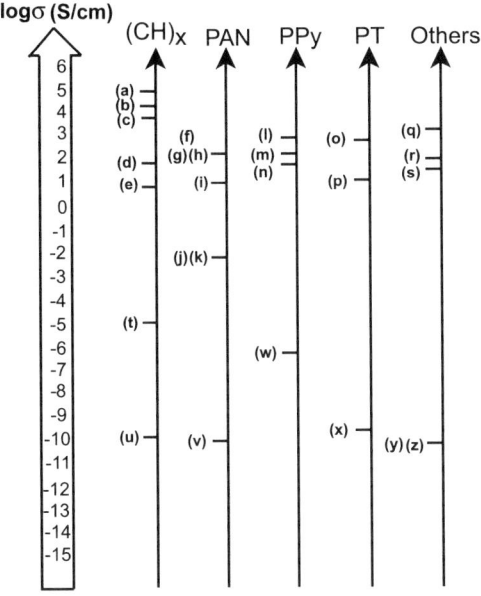

FIG. 1.1. Observed conductivity of several conducting polymers. Low values of conductivity are obtained in the pristine conducting polymers containing high degree of disorder. Conductivity increases with stretching, which aligns the molecular chains of the fiber and produces more ordered regions. The high conductivity under the heading 'Others' is for the sulphuric doped PPV and AsF_5 doped PPP. For more details see Ref. [12] from where the figure has been taken. Note that conductivity close to that of copper has been achieved.

Though conductivity of polyaniline is not as high as that of some other polymers, it is emerging as the material of choice for many applications. It is stable in air and its electronic properties can be easily tailored. It is one of the oldest synthetic polymers, and probably it is the cheapest conducting polymer used in devices. It can be easily fabricated as thin films or patterned surfaces. Polyaniline will never replace the materials which have extremely high conductivity. However, it will be useful for certain specific applications. Andy Monkman has a program to extrude the polymer braids and lay the insulation of the coaxial cables in a single step. The work is supported by a cable company [3]. Properties of polyacetylene are discussed in detail in Chapter 2.

Excitations in conducting polymers consist of localized polarons. The polarons can be singly charged, doubly charged or neutral polaron-exciton [15]. This localization results in large binding energy of the excitons. If there are no quenching centers in the polymer, this localization results in very high luminescence efficiencies. There are no satisfactory theories of conducting polymers. The widely quoted SSH Hamiltonian neglects electron energies. It gives results, which agree with experiments in many cases. Many experiments show discrepancy with the theory by up to 50%. Almost any theory can be defended by using a restricted set of data. In many cases the inconvenient data is dismissed as the product of bad samples by simply ignoring the data altogether. Optical properties are discussed briefly in Chapter 3. More detailed description of individual polymers is given in the chapters on LEDs and solar cells.

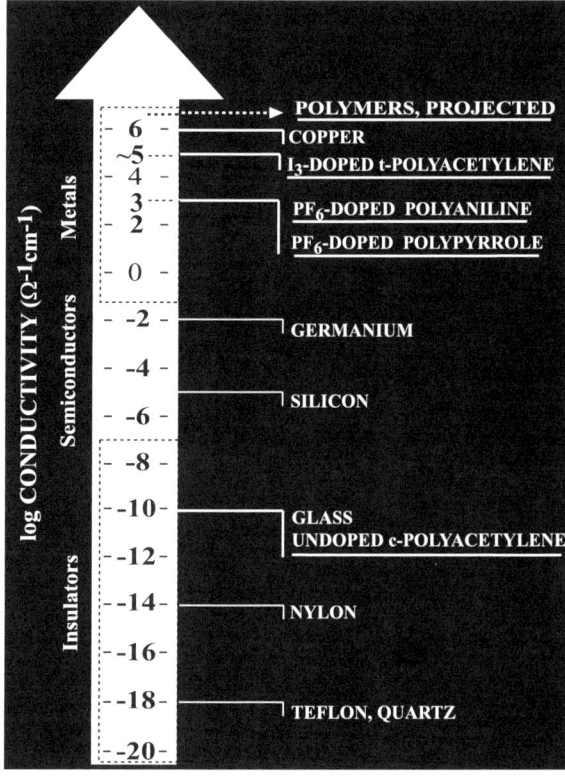

FIG. 1.2. Conductivity of polymers varies over 16 orders of magnitude. Both experiments and theory suggest that ultimate conductivity will be larger than that of copper as shown by the dotted arrow [12,13].

Much of the modeling of devices requires knowledge of the mechanism of the transport of charge carriers in the conducting organics. Because of the importance of transport properties, extensive work has been done on this topic. Transport properties are also discussed in detail in Chapter 3.

1.4. Active Devices

Since the discovery of electroluminescence from organic materials it has been recognized that the conducting polymers are important for fabricating the Organic Light Emitting Diodes (OLEDs). To improve the performance of OLEDs a good understanding of basic device physics is very necessary. Today OLED technology has become a competitor for conventional light sources and display technologies like liquid crystal displays. Displays based on organic semiconductors have already entered the market. Flat panel displays and LEDs, antistatic coatings, electromagnetic shielding, lights for toys, microwave ovens are important applications being pursued now. UNIAX Corporation in Santa Barbara has demonstrated alphanumeric OLED displays [3].

There is a large demand of green, blue and violet Light Emitting Diodes (LEDs) and lasers. These light sources are required for full color displays, laser printers, high-density information storage and for under-water optical communication. By reducing the laser wavelength from the present red to 360 nm the information that can be stored in a compact disc can be increased by a factor 4. These emitters will also be very useful in designing and developing instruments for medical diagnostics and all solid state full color flat panel displays. They will save power by replacing incandescent lamps with LEDs. The quest for blue and green semiconductor light sources has been on for over 30 years. Due to several breakthroughs in the early 1990s, violet, blue and green semiconductor light emitters have become available. During the present decade the evolution and rise of blue and green LEDs and lasers have taken place at an extraordinary pace. The materials and optoelectronic devices emitting light are reviewed in Chapter 4. One of the most spectacular achievements of this decade is the development of the violet, blue and green light emitting diodes, optically pumped lasers and laser diodes. These devices are based on conducting polymers and oligomers, III-Nitrides, and II-VI semiconductors. Work on the properties of organic materials and devices is described in Chapter 4. Organic electroluminescence has bright feature for low cost, lightweight, large area flexible full color flat panel electronic displays. LEDs are discussed in Chapter 4.

Organic solar cells have reached efficiencies exceeding 4%. In fact power conversion efficiencies of organic solar cells have reached an impressive 5%. This has been possible because of the discovery of bulk heterojunction solar cells. Solar cells are discussed in Chapter 5.

First organic thin film transistors were fabricated more than 15 years ago [16]. Very significant improvements have been made in the performance of OFETs. Organic Thin Film Field Effect Transistors (OFETs or OTFTs) are of great interest for both academic and industrial institutions. Several authors have fabricated organic OFETs with performance comparable to the best amorphous silicon transistors. The mobility of the charge carriers is improved considerably. In pentacene transistors mobility of more than 1 $cm^2/V\,s$ has been obtained. The transistors have application as drivers for flat-panel displays, smart cards, electronic barcodes and in other low cost electronic devices. OFETs are discussed in Chapter 6. The transport of the charge carrier in OFETs has been generally interpreted using the mobility model. The mobility of highly pure and defect free crystals of small molecules follows the same behavior as the inorganic semiconductors. In this case, mobility decreases with temperature. The mobility in OFETs is also discussed in Chapter 6.

CHAPTER 2
POLYACETYLENE

2.1. Structure, Growth, and Properties

2.1.1. Structure

Polyacetylene is the simplest conducting polymer. The two forms of conducting polyacetylene, known as *trans*-polyacetylene (or t-PA) and *cis*-polyacetylene (or c-PA), exist. They are shown in Fig. 2.1(a). The bandgap of t-PA is about 1.4 eV and that of c-PA is 2.0 eV. The t-PA is of greater academic and technological interest. It has two degenerate ground states as shown in Figs. 2.1(b) and 2.1(c). The degeneracy of the ground state of t-PA plays a significant role in determining the properties of the polymer. Typical polymers with non-degenerate ground states are shown in Table 2.1.

2.1.2. Growth and Doping of Polyacetylene

The most commonly used method to synthesize the PA is the Shirakawa method. In this method a smooth surface wetted by the Ziegler–Natta catalyst is exposed to the acetylene gas. A film of PA (generally c-PA) is produced on the smooth surface. The c-PA is converted to the t-PA by heating. The process of doping also converts the c-PA to the t-PA.

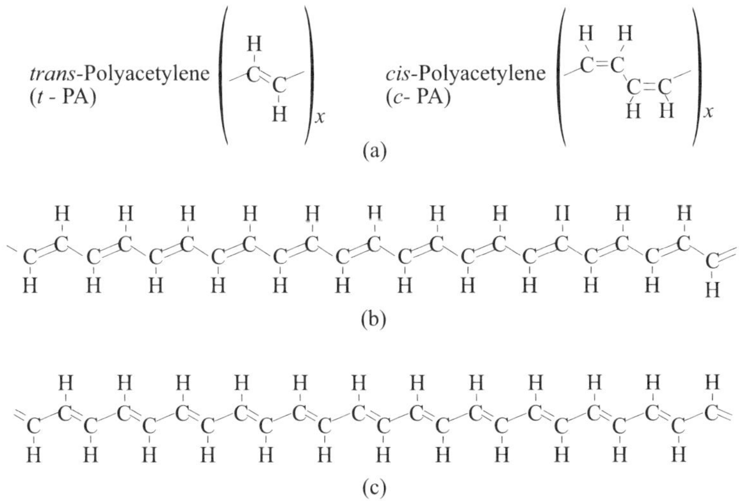

FIG. 2.1. (a) Structure of the t-PA and c-PA repeat units; (b) and (c): two degenerate ground state configurations of the t-PA.

TABLE 2.1
STRUCTURE OF THE REPEAT UNITS AND OPTICAL ABSORPTION EDGE OF FIVE NON-DEGENERATE
GROUND STATE (NDGS) CONDUCTING POLYMERS [17]

Polymer	Structure	Absorption edge
Polypyrrole (PPy)		2.5 eV
Polythiophene (PT)		2.0 eV
Poly(p-phenylene) (PPP)		3.0 eV
Poly(phenylene vinylene) (PPV)		2.4 eV
Polyaniline (PANI) emeraldine form		1.6 eV

The t-PA is the high temperature stable form. The films produced by this method are up to 90% crystalline. However they consist of fibrils which fill only about 30% of the volume. The fibrils are 5–50 nm in diameter. The coherence length of the polymer is 10–20 nm due to the isolated chain defects or small changes in the unit cell parameters. The t-PA films have also been prepared by using suitable precursor polymers which can be easily produced as a film from the solution. On heating the film, volatile components evaporate and the film of the desired polymer is formed. The films can be oriented by stretching them during the transformation process. A third method to deposit the polymer films is electrochemical. The monomers of the desired polymer are dissolved in an appropriate solution and a potential is applied to the two electrodes dipped in the solution. This method is frequently used for depositing the polypyrrole films. These films can also be oriented by stretching. The stretching can be done by a factor up to 15. In the stretched films the conductivity anisotropy can be as high as 200.

Alkali metals are commonly used as donors. The acceptors generally used are I_3^-, AsF_5^-, ClO_4^- and $FeCl_4^-$. The doping is achieved by exposing the polymer to a vapor or liquid containing the desired dopant. Since the dopant ions are highly electropositive or electronegative, doping can also be achieved by using the polymer film as an electrode in an electrochemical cell containing the dopants in the solution. The electrochemical method of doping provides a better control over the dopant concentration. At low concentrations the dopants are distributed randomly. For larger dopant concentrations the polymer has two phases. One phase is dopant rich and in this phase the dopant ions are distributed in a lattice structure. The other phase has low dopant concentration and the distribution of dopant ions is random. At very high dopant concentration transition from insulator to metal takes place.

As mentioned in the first chapter, monolayer control of thickness of organic films has been obtained using the Langmuir–Blodgett film deposition technique and layers grown by self-assembled monolayer from solutions [11]. Since its emergence in the mid-1980s, Organic Molecular Beam Epitaxy (OMBE) has become an important technique for deposition of organic films with monolayer thickness control in atomically clean environments. Extreme chemical and structural controlled films are obtained.

2.2. Band structure of t-PA

In the *trans*-polyacetylene (t-PA) the neighboring sp^2 hybridized orbitals pointing at each other between two C atoms form bonding and antibonding states with a separation of about 20 eV. The bonding energy levels are completely filled by the two electrons donated by the two C atoms. These electrons, known as σ electrons, contribute to the rigidity of the polymer. In most organic materials the difference in the lengths of double and single bonds is 0.2 Å. In the t-PA the difference is much less, 0.07 or 0.08 Å. However it is useful to think of double and single bonds in the t-PA also. Altogether a C atom gives three of its 4 valence electrons to the σ bonds. The fourth electron occupying the p$_z$ orbitals couples with the neighboring p$_z$ orbitals forming the π band. In the tight binding model the coupling energy t is given by [14],

$$t = -\int \phi_n^*(r+1) H \phi_n(r) \, dr, \tag{2.1}$$

where $\phi_n(r)$ and $\phi_n^*(r+1)$ are the orbitals in the nearest neighboring positions and H is the Hamiltonian. The integral in Eq. (2.1) is known as the resonance or the transfer integral. For uniform spacing of the p$_z$ orbitals as shown in Fig. 2.2(a) (and neglecting the zigzag), the value of t is denoted by t_0. A reliable calculation of the energies t or t_0 is difficult. The energies are usually determined by fitting the theory with the experiment.

For uniform spacing shown in Fig. 2.2(a) the energy dispersion relation for the π band is [14]

$$\varepsilon_k = -2t_0 \cos ka. \tag{2.2}$$

Here k is the wave vector of an electron in the π band and a is the spacing between the nearest C atoms on the polymer backbone. The energy dispersion given by Eq. (2.2) is shown by the dashed line in Fig. 2.2(c). Since each C atom gives only one electron, the π band is half filled and the t-PA should be a metal. However as was first shown by Pieirls, degeneracy of the band in the one dimensional (1D) metals is removed by spontaneous distortion of the 1D chain. In the case of t-PA the distortion consists of dimerization shown in Fig. 2.2(b). Dimerization opens up a gap ($\sim$1.4 eV) and the t-PA becomes a semiconductor. The band structure of the dimerized chain is shown by the solid lines in Fig. 2.2(c).

The band structure of the dimerized chain is shown by the solid curves in Fig. 2.2(c). A gap in the band opens and the polymer now becomes a 1D semiconductor. The bandgap

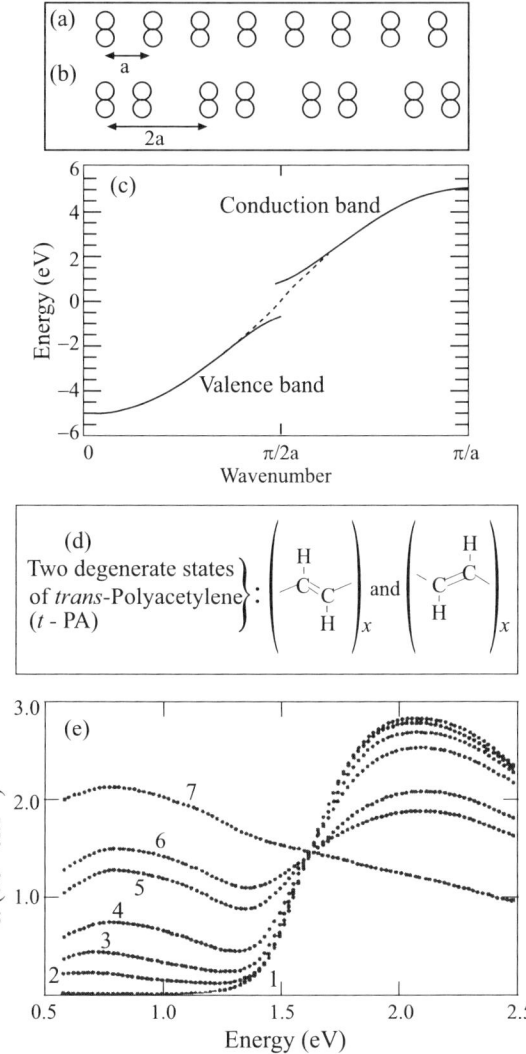

FIG. 2.2. (a) The p_z orbitals of undimerized t-PA chain, (b) the orbitals after dimerization, and (c) energy versus k for the π band before dimerization (dashed line) and after dimerization (solid line) [14]. (d) Two degenerate structures of the t-PA are shown. (e) Optical absorption spectra of undoped (curve 1) and doped t-PA [18]. The doping increases with the curve numbers.

in this case is 1.4 eV. The structures of the basic unit C_2H_2 of t-PA are shown in Fig. 2.2(d). There are two configurations which give the same ground state energy. The two configurations can occur in the same chain which gives rise to the formation of polarons, solitons and energy levels in the bandgap. Measured optical absorption of t-PA is shown in Fig. 2.2(e). The width of the conduction band is small. The absorption at ~0.7–0.8 eV is due to the formation of soliton midgap levels.

Distortion of the chain leading to the dimerization is induced by the electron–phonon interaction. The interaction arises due to the dependence of the transfer integral (given by Eq. (2.1)) on the spacing between the atoms which changes due to the vibrations. Since the displacement u_n of the nth C atom due to the dimerization is small, the transfer integral for the dimerized chain can be expressed as

$$t = t_0 + \alpha(u_n - u_{n+1}), \quad \alpha = (\partial t/\partial u)_0. \tag{2.3}$$

To calculate the distortion due to the Pieirls instability and determine the bandgap Hamiltonian including the electron–phonon interaction is diagonalized and its eigenvalues are calculated. The electron–electron interaction is neglected. The eigenvalues are given by

$$E_k = \pm\left(\varepsilon_k^2 + \Delta_0^2\right)^{1/2}, \quad \Delta_0 = 4\alpha u_0, \tag{2.4}$$

where u_0 is a constant and ε_k is given by Eq. (2.2). The Pieirls gap is $8\alpha u_0$. The value of u_0 is determined by minimizing the total energy (the sum of the electronic energy and the lattice distortion energy). The final expression for half the gap Δ_0 can be written in the form:

$$\Delta_0 = (8t_0/e)\exp(-1/2\lambda), \quad \lambda = 2\alpha^2/\pi K t_0. \tag{2.5}$$

Here K is the stiffness constant. The values of the parameters generally used in Eq. (2.5) are given in Table 2.2. These values lead to $\alpha = 4.2$ eV/Å and the electron–phonon coupling constant $\lambda = 0.19$. However using Eq. (2.5), the value of Δ_0 comes out to be only 0.53 eV which is considerably smaller than the experimental value of 0.7 eV. Eq. (2.5) is approximate because the electron–electron interactions have been neglected. Also the optical absorption experiments do not give a reliable value of the gap because the absorption can occur by processes other than electronic excitation.

Takayama et al. [19] used a continuum form of the Hamiltonian and derived the following value of Δ_0

$$\Delta_0 = 4t_0\exp(-1/2\lambda). \tag{2.6}$$

This equation yields $\Delta_0 = 0.7$ eV which is in better agreement with the experiment.

Several modifications have been made in the theory. These modifications have been reviewed in Ref. [7, see pp. 825–913].

TABLE 2.2
VALUES OF PARAMETERS GENERALLY
USED FOR CALCULATING THE PIEIRLS
GAP IN THE t-PA [14]

Parameter	Value
$2\Delta_0$	1.4 eV
u_0	0.04 Å
t_0	2.5 eV
K	21 eV/Å^2

2.3. The Solitons and the Polarons

2.3.1. THE SOLITONS

The t-PA is degenerate in the ground state. The two configurations shown in Figs. 2.1(b) and 2.1(c) have the same energy. The two configurations can occur on the same chain, the transition from one to another takes place by distortion of bonds extending to some 14 sites shown in Figs. 2.3(a) and 2.3(b). The distortion consists of changes in electronic charge in the bonds and in the bond lengths. Conducting polymers are quasi-one dimensional (quasi-1D) and deformation of the chain occurs relatively easily. The dot at the center shows an electron still bound to the chain but with a different energy which lies close to midgap. The displacement of the C–H groups due to the distortion is shown in Fig. 2.3(c). The calculated values of the bond lengths in the dimerized undistorted chain are: $r_{C-C} = 1.346$ Å and $r_{C-C} = 1.446$ Å, which give the difference between the two bond lengths $u_0 = 0.10$ Å. Experimental values vary between 0.08 and 0.1 Å [7]. The electron along with the distortion of the chain constitutes the soliton in the t-PA. Two configurations of the soliton, designated as S and $\tilde{S}$ exist. They are shown in Figs. 2.3(a) and 2.3(b). In the undoped polymer the soliton is neutral. However since it has an electron, it has a spin 1/2 and obeys the Fermi–Dirac statistics. The soliton accepts an electron and is negatively charged to form (S^-) if the polymer is doped with donors. In the p-type polymer, the soliton looses an electron and becomes positively charged. The

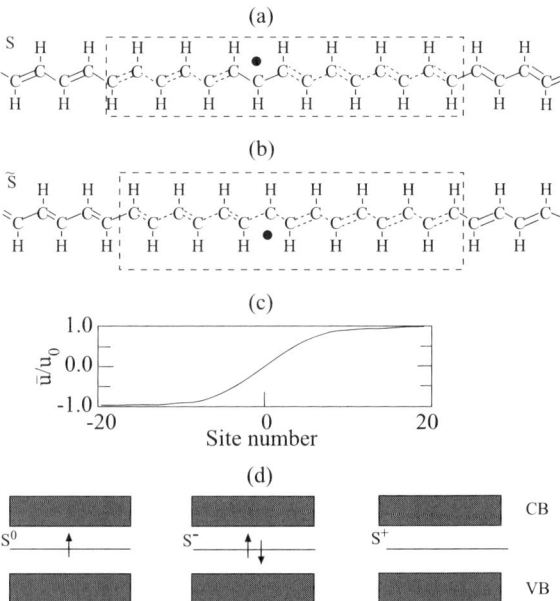

FIG. 2.3. (a) and (b): Structure of the soliton S and antisoliton $\tilde{S}$ (the filled circles represent the electrons still bound to the polymer chain), (c) normalized displacement in the C–H positions due to formation of the soliton, and (d) energy levels of the neutral and the charged solitons (the arrows show the electrons with spins up or spins down). The figure is adapted from different figures given in Ref. [14].

electronic structure of the energy levels of the neutral and charged solitons are shown in Fig. 2.3(d). The charged solitons drift under the action of applied field and give rise to an electric current. The charged solitons which conduct electricity are spinless, a behavior strikingly different from the charge carriers in other semiconductors. They obey the Bose–Einstein statistics. Addition of an electron to the polymer which is nondegenerate in the ground state does not create a soliton, it rather creates a polaron discussed in the next subsection. Solitons can be formed only in those polymers which are degenerate in the ground state. The *t*-PA is one of the few such polymers.

The energy E_s needed to create a soliton consists of the electronic energy and the lattice distortion energy. In the continuum model the energy is given by

$$E_s = 2\Delta_0/\pi. \tag{2.7}$$

Assuming the chain length to be much larger than the displacements, the kinetic energy of the soliton comes out to be

$$KE = \left(\frac{4Mu_0^2}{6a\zeta}\right)v^2 \tag{2.8}$$

where M is the mass of the CH group and 2ζ is the length of the polymer chain. The apparent mass of the soliton is given by

$$m_s^0 = 4Mu_0^2/3a\zeta. \tag{2.9}$$

The superscript ^{0} indicates that the mass is in the low velocity limit. The value of m_s^0 comes out to be $\sim 5m_e$ where m_e is the mass of the free electron.

2.3.2. THE POLARONS

When an electron is added (by doping with donors or by photoexcitation) to the NDGS semiconductor polymer, it does not go into the conduction band as is the case in a conventional semiconductor. It deforms the polymer chain as shown in Fig. 2.4. The actual

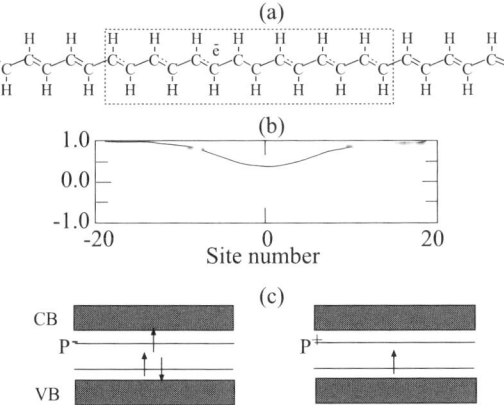

FIG. 2.4. (a) Structure of the negatively charged polaron, (b) normalized displacement of the CH groups in the polaron, and (c) electronic structure of the negatively and positively charged polarons (the arrows show the electrons with spins up or spins down). The figure is constructed using different figures given in Ref. [14].

deformation occurs over some 20 sites. In the figure the deformation is shown to occur over fewer sites for convenience. The displacement of the C–H groups due to this distortion is shown in Fig. 2.4(b). The electron and the deformed chain constitute the polaron. The deformation creates the energy levels in the gap shown in Fig. 2.4(c). The electron falls in the level close to the conduction band edge. The electron along with the deformed configuration of the chain is the negatively charged polaron P^-. An analogous configuration exists for a positive polaron in a polymer doped with p-type impurities. The positive polaron is also shown in Fig. 2.4. The electron (or the hole if the polaron is positive) along with the deformation diffuses, and under the action of an applied field, it drifts causing a current to flow. Thus the charge carriers in the NDGS polymers are polarons (or bipolarons discussed below). If an electron and a hole are created simultaneously, both negative and positive polarons can be formed in the same polymer. When these two polarons meet they recombine and are annihilated.

If two polarons of like sign are formed close to each other, a bipolaron is formed. Two energy levels are created by a bipolaron in the bandgap. They are both occupied either by two electrons (for a negative bipolaron) or by two holes, i.e. they are empty (in the case of a positive bipolaron). The bipolaron has no spin. The bipolaron may not be stable because of the repulsion of the two polarons which constitute the bipolaron. However the dopant ions in the neighborhood stabilize the bipolaron.

Feldblum et al. [18] and others (see references given in the review [8]) have studied the optical absorption in the t-PA, undoped and doped electrochemically with different concentrations of the ClO_4^- ions. Curve 1 in Fig. 2.2(e) shows the absorption for the undoped polymer. The position of the absorption edge indicates that the polymer is a semiconductor with a bandgap of about 1.4 eV. On doping the polymer an additional peak approximately at midgap (at 0.7–0.8 eV) is observed. The height of this peak increases with doping concentration. This peak is attributed to the midgap level created by the soliton. The position of the peak is independent of the nature of the dopant, i.e. it is observed for a variety of donors and acceptors. This confirms that the peak is not due to the absorption directly associated with the dopant but it is due to the soliton energy levels.

Careful experiments have not revealed any significant midgap absorption in the undoped polymer. It has been suggested that in the undoped polymer the energy levels due to solitons are close to the band edges. The simple theory which predicts midgap levels in the undoped polymer does not take into account the electron–electron interactions and is very approximate.

Extensive work has been done on the photogeneration of defects near the midgap in the t-PA. A good review of this work is given in Ref. [8]. We give a summary of the important results here. Transient spectroscopy shows that absorption due to photogenerated gap state gives two peaks, one at 1.4 eV due to band-to-band transitions and another at 0.5 eV due to midgap states. The band-to-band transition is observed in the undoped t-PA also and the midgap states arise only on irradiation or on doping. The midgap state absorption decays to half of its value in a few milliseconds whereas the band-to-band absorption decays more rapidly, in about 100 μs. Photoemission from the midgap states has also been observed. The midgap states are produced in less than 10^{-13} s and the midgap excitations are highly mobile. Typical results of photoinduced changes in transmission are shown in Fig. 2.5.

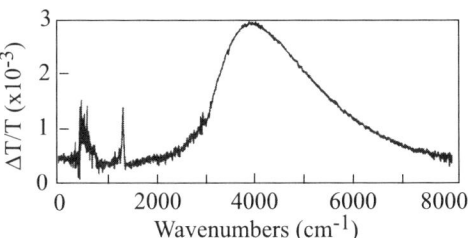

FIG. 2.5. Photoinduced changes in optical transmission in the *t*-PA. The IR vibrational bands are at 1370, 1260, and 500 cm^{-1} [8,20,21].

The 0.5 eV midgap state absorption band is quite prominent but is rather broad. The difference in the midgap energy is 0.7 eV and the absorption peak at 0.5 eV is due to the effect of the electron–electron interaction on the transitions [8]. Since the initial state involves two electrons in the midgap level and the final state involves only one, the Coulomb interaction lowers the energy of transition. The 0.5 eV band is quite similar to the midgap absorption in the doped *t*-PA shown in Fig. 2.2. This demonstrates that this band is associated with the same center in both the photoirradiated and the doped polymers.

Three IR bands at 1370, 1260, and 500 cm^{-1} are also seen in Fig. 2.5. These energies are in the gap between the Raman mode of pure *t*-PA and are predicted by theory for the photoinduced solitons. The band near 500 cm^{-1} is attributed to the pinned soliton–antisoliton pair. The IR vibrational bands are also observed at ~900, 1270, and 1370 cm^{-1} in the doped crystals. The band at 1270 cm^{-1} is rather weak. One possible explanation of the existence of the IR bands is that the coupling of the π electrons with the C-atom vibration along the chain increases the dipole-moment of the C vibrations making them infra-red active. The soliton vibrations in the Coulomb field of the impurity can also give rise to the infra-red active modes [14]. A more detailed discussion of these bands is given in [6].

The photoinduced 0.5 eV band is strongest at low temperatures and starts decreasing in strength at 150 K. It becomes unobservable at 250 K. Photoconductivity is maximum at room temperature and becomes very weak below 200 K. This observation leads to the conclusion that the charged solitons are free to move only above 150 K.

There has been considerable discussion on the mechanism of generation of soliton–antisoliton pair by incident photons. Early theory indicated that the process is indirect [8,6]. A measurement of the excitation profile for the pair photoproduction has been used to determine the mechanism. A *t*-PA sample was irradiated with photons of different energies and production of the soliton–antisoliton pairs was monitored by measuring the transmission at 1370 cm^{-1}. The change in transmission at this energy is a measure of the number of the pairs. The results of the measurements are shown by the filled circles in Fig. 2.6. The band-to-band absorption coefficient is also plotted in the figure. The threshold of the band-to-band absorption is at about 1.4 eV as expected. The important observation is that the threshold for the photoproduction of the solitons is at about 1 eV, considerably below the threshold for the band-to-band absorption. This energy agrees with Eq. (2.7) if bandgap is taken to be 1.6 eV. The number of solitons increases exponentially with the energy of the incident photons up to about the threshold of the band-to-band

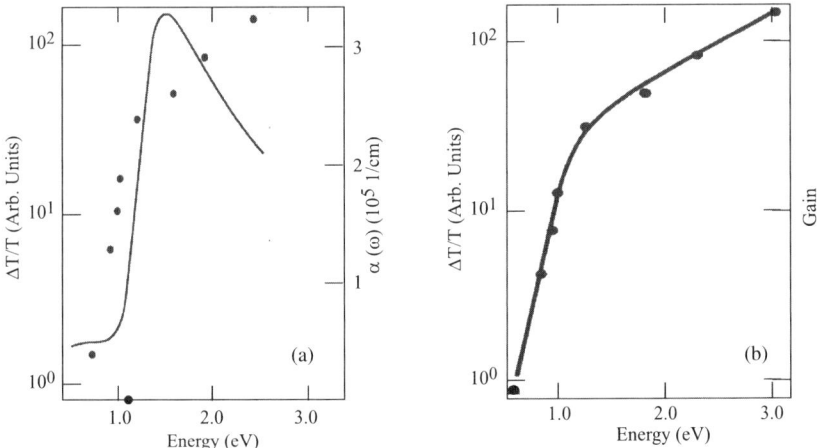

FIG. 2.6. (a) Photoinduced changes in transmission at 1370 cm^{-1} versus energy of the photons are shown by the filled circles. The solid curve shows the optical absorption versus energy of the photons [21]. (b) The filled circles show the same excitation profile as shown in (a). The photoconductivity rate of Ref. [22] is shown by the solid curve. The figure is taken from Ref. [8].

absorption. At the threshold the production rate becomes slower, it again increases exponentially but with a smaller slope. This shows that below the threshold for the absorption, the soliton pairs are produced directly. Above this threshold the production is via the electron–hole pairs.

2.4. Transport Properties

2.4.1. MOBILITY IN SELECTED POLYMERS

Mobility has been studied extensively in the pristine conducting polymers. Measurements have been made using the time of flight method, by measuring electroluminescence in a light emitting diode (LED), by studying the drain current in a thin film transistor in which the polymer is used for the active channel (the method is designated as the Field Effect or the FE method), by measuring the diffusion coefficient (using magnetic resonance methods) and converting it to the mobility using Einstein's relation, by measuring space charge limited currents and by measuring transient photoconductivity. The transport of the charge carriers takes place by intra-chain hopping, by inter-chain hopping and also by drift. At very high dopings, insulator to metallic transition takes place. Several sophisticated theoretical models have been advanced to interpret the observed values of the mobility. In particular Mott's theory of variable range hopping, mobility edge, and insulator-to-metal transition have been used. The interpretation of the results becomes complicated because generally the polymer contains mixture of metallic and insulating regions. The degrees of disorder, trapping and detrapping also play a significant role in determining the mobility. The experimental values of the mobility are highly sensitive to the method of preparation of the polymer and to the heat treatment. We give here experimental results obtained by the time of flight method, by the FE method and by the

TABLE 2.3
TIME OF FLIGHT HOLE MOBILITIES IN THREE CONDUCTING POLYMERS [17]

Material	Mobility μ (cm^2 V^{-1} s^{-1})	Measurement field E, E-dependence (V cm^{-1})
PPV	10^{-4}–10^{-3}	Independent of E
	$285 < T < 385$ K	$10^4 < E < 2 \times 10^5$
PPV	$10^{-4} < \mu < 10^{-3}$	10^4
PPV	3×10^{-3}	1.5×10^3
PPV	$\sim 5 \times 10^{-6} < \mu < 5 \times 10^{-5}$	4.5×10^4
PPV	$10^{-7} < \mu < 3 \times 10^{-6}$	$\ln \mu \propto E^{1/2}$, $10^5 < E < 10^6$
PPV	$\sim 10^{-6}$	2×10^4
PPPV	$10^{-3} < \mu < 10^{-4}$	$\ln \mu \propto E^{-0.85}$
		$3 \times 10^4 < E < 3 \times 10^5$
α-6T	5×10^{-6}	$3 \times 10^5 < E < 8 \times 10^5$

diffusion method. Some comments on the magnitude of the observed mobility and on the field and temperature dependence of the mobility will be made. A discussion of the theoretical models and their comparison with the experimental results will not be attempted.

In the time of flight method, light is absorbed at one surface of the polymer having a transparent contact. The light with a large absorption coefficient is chosen so that it is absorbed over a small thickness. The time taken by the pulse of carriers so generated to travel to the other end is measured and the mobility is determined. Unfortunately the method does not yield reliable values of the mobility when trapping and detrapping of the carriers take place. If a voltage is applied to a LED, the luminescence occurs after elapse of a short but finite time. This time is needed for the carrier to travel to the opposite carrier and recombine. The measurement of this time also yields a value of the mobility. The values of mobility μ obtained by the conventional time of flight method are shown in Table 2.3. The measurements on oligomer sexithiophene (α-6T) and poly(p-phenylene vinylene) (PPV) were made by the luminescence method. The several values of PPV given in the table are taken from different authors. The values do not agree with each other. The PPV samples were made by heating a non-conducting polymer and converting it to the conducting polymer. The heat treatment has a large effect on the disorder and the conjugation length and therefore on the mobility. The conjugation length increases with heating and approaches a limiting value of about 7 monomers. It is not clear whether the mobility is only due to hopping. In some cases the chains were long and the mobility might have had a drift component also.

It is seen from Table 2.3 that except for the results given in the first row, the transit time is dispersive, i.e. the mobility depends on the applied field. In some cases the field dependence could be expressed as $\mu = \exp(BE^{1/2}/kT)$ and temperature dependence as $\mu = \exp(-E/kT)$, E being in the range from 0.1 to 0.15 eV. For α-6T the value 5×10^{-6} cm^2 V^{-1} s^{-1} of the mobility determined by the time of flight method is a few orders of magnitude smaller than the value determined by the FE method shown later in Table 2.4.

Most measurements by the FE method have been made in the hole accumulation layer formed in the channel on application of a voltage between the gate and the source. Drain

TABLE 2.4
FIELD EFFECT HOLE MOBILITIES IN SEVERAL CONDUCTING POLYMERS [17]

Material	μ_{FE} (cm^2 V^{-1} s^{-1})
Sexithiophene	4.3×10^{-1}
Sexithiophene	2×10^{-3}
Sexithiophene	3×10^{-2}
Dihexylsexithiophene	7×10^{-2}
Dimethylsexithiophene	2×10^{-2}
Quaterthiophene	2×10^{-7}
Quinquethiophene	10^{-5}
Octithiophene	2×10^{-4}
Diethylquaterthiophene	9×10^{-5}
Diethylquinquethiophene	9×10^{-4}
Polythiophene	10^{-5}
Polythiophene	2×10^{-4}
Poly(3-hexylthiophene)	10^{-4}–10^{-5}
Poly(3-hexylthiophene)-based Langmuir–Blodgett film	7×10^{-7}
Quinquethiophene-based Langmuir–Blodgett film	1×10^{-5}
Polythienylene vinylene	2.2×10^{-1}
trans-polyacetylene	$<10^{-5}$–6.5×10^{-4} depending on carrier concentration

current I_d is measured and using the known values of the dimensions of the channel, the mobility is determined. The mobilities determined in this manner are shown in Table 2.4. Table 2.4 shows that sexithiophene and its dihexyl derivative have the highest mobility. X-ray measurements of these polymers show that they are highly ordered. The mobility also increases in going from quaterthiophene to quinquethiophene to sexithiophene. Presumably this is due to the increase in the conjugation length in going from quater- to sexithiophene through quinquethiophene. However the mobility in octithiophene is smaller than sexithiophene, probably due to defects present in this polymer.

A comparison of the mobility values given in Figs. 2.3 and 2.4 shows that the mobility for α-6T determined in the thin film by the FE method is much larger than the mobility in the bulk determined by the time of flight method. The trapping is more important in the time of flight method than in the FE method. In the FE method there is a steady state 'large' concentration of holes. In the absence of holes, the traps (with energy levels below the Fermi level) contain electrons and are efficient trapping centers for the holes. If the hole concentration is large, as is the case in the FE method, all the traps become empty and cannot trap the holes. In the FE method the results should also depend on the properties of the interface between the polymer and the insulator. Mobility in α-6T measured by the FE method increases on annealing at high temperatures. The crystallite size in the unannealed samples is about 100 nm. On annealing it increases to about 5 μm. The increase in the crystallite size increases the mobility.

Measurements of diffusion of neutral solitons have been made using the electron spin resonance (ESR) and the nuclear magnetic resonance (NMR). An unambiguous interpre-

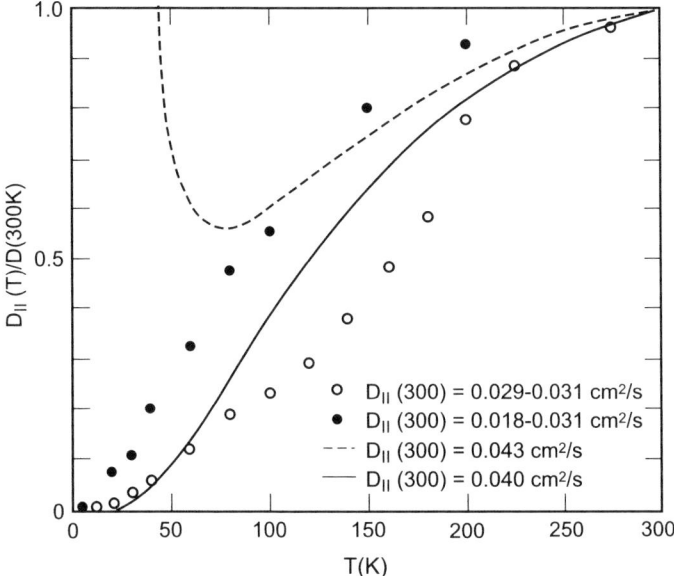

FIG. 2.7. Intrachain diffusion coefficient $D_\parallel$ of solitons in the t-PA versus temperature. The closed circles are the data from Ref. [24] and the open circles from Ref. [26]. The dashed curve shows the results of theoretical calculations including only the phonon scattering and the solid curve shows the calculated value including 0.01 eV barriers [23].

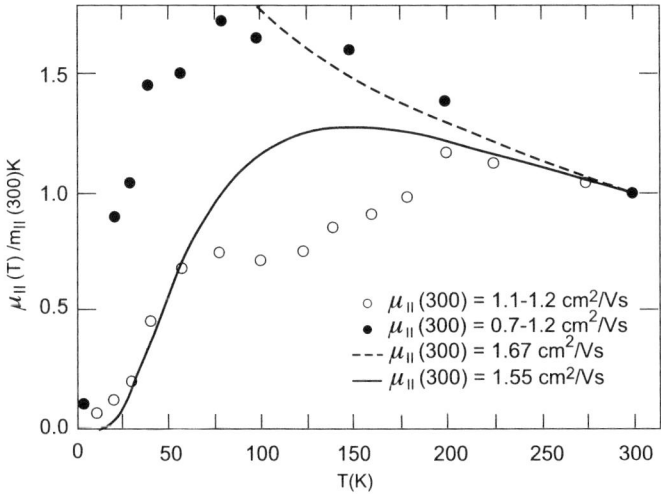

FIG. 2.8. Intrachain mobility of solitons in the t-PA calculated using the values of the diffusion coefficient given in Fig. 2.7 and the Einstein relation given in Eq. (2.10). Labeling of the curves and the symbols are the same as in Fig. 2.7 [25].

tation of the experiments is not easy. The difficulty arises because a fraction of the solitons is trapped. This fraction varies with temperature and its effect of the magnetic resonance experiments can be determined only if the experiments are performed with great care. The results of such careful experiments [23] are shown in Fig. 2.7. The diffusion coefficient decreases slowly with temperature from 300 to 200 K. As the temperature decreases further, the diffusion coefficient decreases fairly rapidly. The actual values of the diffusion coefficient are rather low. Any defect on the chain creates a barrier of height V_0 to the diffusion. Efforts have been made to fit the diffusion data to the theory by treating V_0 as an adjustable parameter. The calculated curve for $V_0 = 0.01$ eV [25] is also shown in Fig. 2.7.

The mobility calculated using the measured value of $D_\parallel$ and the Einstein relation [25]

$$D_\parallel = \mu_\parallel kT/|e|, \tag{2.10}$$

is plotted in Fig. 2.8. The mobility obtained by this method is essentially the drift mobility.

2.4.2. CONDUCTIVITY AND SUSCEPTIBILITY

Fig. 2.9 shows that in highly oriented and heavily doped t-PA conductivity values in excess of 10^5 Ω^{-1} cm^{-1} can be obtained. The samples were prepared with care and stretched by a factor 5 so that they were highly oriented. The high conductivity suggests that they are metallic. The metallic nature was confirmed by the 100% reflectivity observed with similar samples. The conductivity is highly anisotropic. With a draw ratio of 15, the ratio $\sigma_\parallel/\sigma_\perp$ reached value 250. Absorption of light polarized perpendicular to the chains is also small. These observations suggest that the polymers are semiconductors

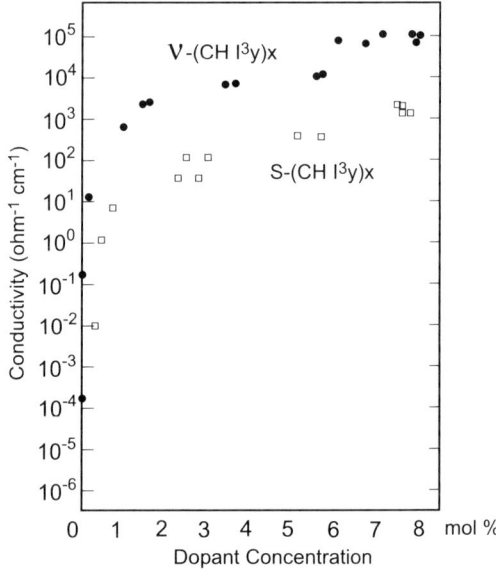

FIG. 2.9. Conductivity versus dopant concentration for two iodine doped t-PA samples. (S): a sample fabricated by the Shirakawa method and (v): samples fabricated by the Tsukamoto method [27].

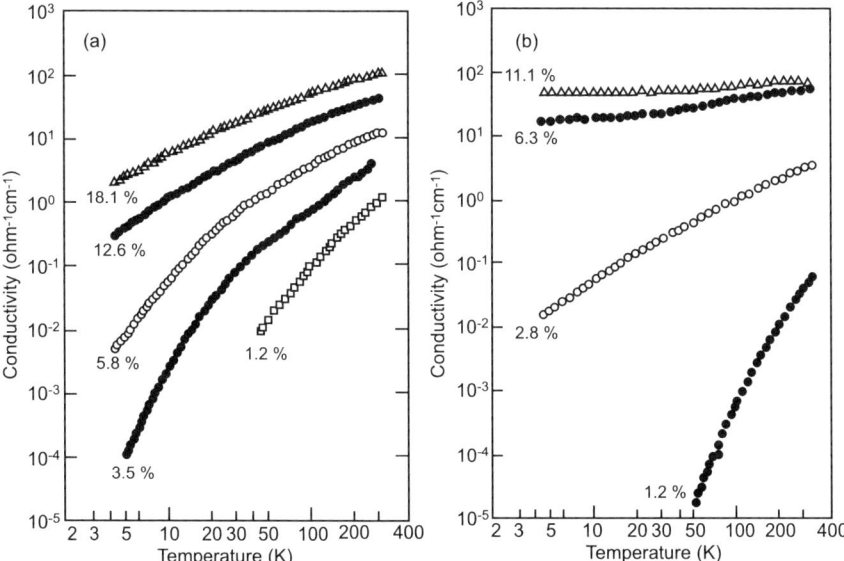

FIG. 2.10. DC conductivity versus temperature plots for (a) iodine doped and (b) FeCl$_3$ doped t-PA [28].

in the perpendicular direction. If that is true, the ratio $\sigma_\parallel/\sigma_\perp$ should reach a value equal to unity at low doping concentrations. It is also seen that there is a step in the conductivity plots at about 6% impurity concentration. It has been suggested that the structure of the impurity distribution or of the polymer changes at this impurity concentration. However in conventional semiconductors the transition from insulator to metal occurs quite abruptly at a certain impurity concentration. Such abrupt transition can not be ruled out in the polymers.

The measured conductivity versus temperature plots for the iodine doped and FeCl$_3$ doped t-PA are shown in Fig. 2.10. The figure shows that the conductivity at low dopant concentrations is thermally activated. As the dopant concentration increases, the conductivity becomes more and more metallic. Susceptibility measurements show that the conductivity is by the transport of carriers which have no spins. It has been suggested that the charge carriers are the charged solitons which have no spins. The suggestion is however weak because no convincing explanation is given as to how the soliton overcomes the strong pinning with the donor or the acceptor ions.

The results of photoconductive gain and excitation profile are remarkable. They show complete agreement of the spectral dependencies of photoproduction of solitons and photoconductive gain. They provide further support that the photoproduction of solitons occurs at energy lower than the threshold for the band-to-band absorption. They also show that the charge carriers are the soliton in the t-PA.

We now discuss the experimental evidence that the charged solitons have zero spins. Let N_s be the number of spins and N_{ch} be the number of the charge carriers. If the charge carriers are Fermions, the ratio

$$R = \frac{N_s}{N_{ch}} \qquad (2.11)$$

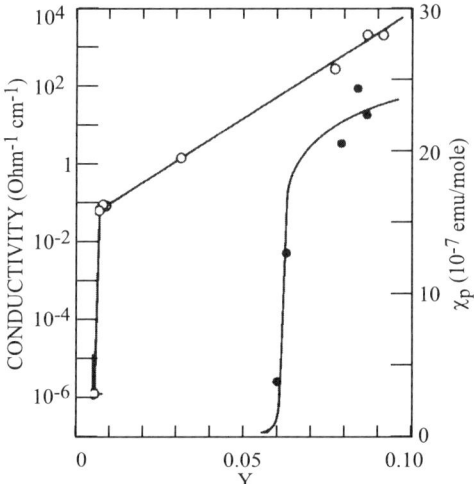

FIG. 2.11. Room temperature conductivity and temperature independent Pauli susceptibility versus dopant concentration Y for $[Na_y^+(CH)^{y-}]_x$ [8].

is equal to unity. If $R = 0$ for the solitons, the spin–charge relation is reversed. Extensive ESR measurements show that in the polymer containing large density of photoinduced solitons, the ratio R is $<10^{-2}$. Calculations also show that the soliton are directly produced as charged entities.

Further evidence that the charged solitons have zero spin comes from the magnetic studies of the doped t-PA. The transition from semiconductor to metal occurs in the polymer when doping exceeds some 5%. Magnetic susceptibility of the polymer remains low throughout the transitional regime from 1 to 5% doping. Midgap transition and the IR modes are observed in this range. This leads to the interpretation that the conduction in this regime is by the unpinned charged solitons. Some experiments showed that the susceptibility increased throughout the regime. To interpret these experiments, it was postulated that the metallic islands were formed continuously beginning at the lower dopant concentration. Final transition occurs at 5% dopant concentration when the islands meet. A percolation model was fitted with the observation successfully.

Later experiments were made on thin films (∼0.2 μm) of the polymers with more uniform concentration of the dopant [8, p. 747]. The simpler nonreactive Na^+ sodium ions were used as dopants. The measured susceptibility χ_p and the conductivity σ_{DC} are shown in Fig. 2.11. The data shown in this figure cannot be fitted with Mott's variable range hopping model, the model gives too small conductivity. The conductivity data in the transitional regime was interpreted as being due to a high density of charged spinless solitons.

We have discussed the physics and technology of polyacetylene in detail in this chapter. A good understanding of the physics of polymers became possible as a result of the work done on polyacetylene.

CHAPTER 3
OPTICAL AND TRANSPORT PROPERTIES

3.1. Effect of Electric Field on Photoluminescence (PL)

Time resolved PL experiments on PPV show that the optical generation of electron–hole pairs is a secondary process. In the primary process excitons are generated by the incident light, the excitons then dissociate into separated electron–hole pairs. Experiments performed in the late 1980s showed that PL in both organic and inorganic semiconductors decreases when an electric field is applied to the semiconductor. In the recent publication [167], Kersting et al. have examined the causes of this reduction. The reduction could occur if the electric field reduces the oscillator strength for emission, e.g. by promoting non-radiative radiation. Alternatively the field may dissociate the excitons into electrons and holes thus reducing the probability of radiative recombination. Kersting et al. [167] have studied the temporal evolution of PL with and without the application of the electric field. The experiments were performed on the 1000 nm thick film of (80% poly(*para*-pyridiyl vinylene)(PPPV) + 20% polycarbonate) polymer blend sandwiched between an ITO/glass and Al electrodes. The main peak of the PL was at 2.51 eV (494 nm) with a vibronic side band at lower energy. The main peak was attributed to $S_1 \rightarrow S_0$ (0–0) transition. On application of 20 V bias the PL intensity decreased by about 20%. Similar results were obtained with a 1 μm thick PPPV film. The results with forward and reverse bias were practically identical. Time resolved PL studies showed that the evolution of PL was ultra fast, the observed time was limited to 300 fs by the time resolution of the setup. Time resolved studies were also made with the application of the electric field on the sample. The electric field had no effect on the rise of PL during the time (100 fs) the incident light is on. The electric field suppresses the PL considerably when the PL starts decaying. In the PL intensity (I_1) versus time t plots (Fig. 3 of Ref. [167]) peak value of I_1 close to $t = 0$ was reduced to about 60% on the application of the electric field. After 100 ps, the quenching reduces the PL to less than 15%. The quenching of the PL is not instantaneous but evolves on a picosecond time scale. The decrease of I_1 can be represented by the following equation,

$$\frac{I_1(E) - I_1(E=0)}{I_1(E=0)} = E^n \tag{3.1}$$

where n has a value larger than 1.

These results show that the main cause of the reduction of PL is the dissociation of the excitons by the electric field and not due to any reduction in the oscillator strength.

Polyacetylene is the most extensively studied polymer and is well understood. However it is not very stable. Most recent work is concentrated on the polymers which consist of chain of rings. Typical polymers which have received considerable attention are shown in

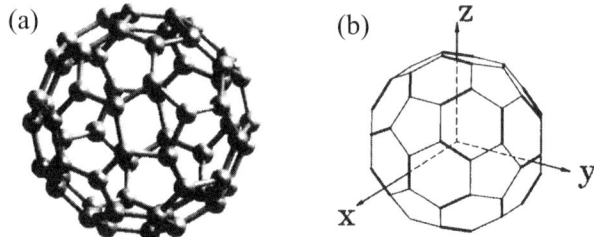

FIG. 3.1. (a) Molecular structure of Buckminsterfullerene C_{60}. The structure on the backside is also visible. The figure is taken from Ref. [1, p. 110]. (b) A view of Buckminsterfullerene C_{60} in one of the two standard forms. In each of these two forms the two fold axes are aligned with cubic (100) axes. The thick lines show the double bonds, see Ref. [1, p. 154].

Fig. 3.2. It is true that degeneracy of the ground state gives the t-PA unusual and interesting properties. The polymers shown in Fig. 3.2 and most other conducting polymers have nondegenerate ground state. However coupling of the excitations with the distortions of the polymer chain is not unique to polyacetylene. It is the general property of the conducting polymers. There are many conjugated polymers which have two ground states which differ in energy not by large amount. Three such polymers with two ground states are shown in Fig. 3.3. The two benzoid and quinoid forms of *para*-phenylene have an energy difference of $\Delta_0 \approx 0.35$ eV per phenyl monomer. The formation of soliton–antisoliton pairs with two neutral domain walls is shown schematically in poly(*para*-phenylene) in Fig. 3.4. The formation energy of the pair is given by,

$$E_{\text{tot}} = 2E_{\text{s}} + n\Delta E_0 \tag{3.2}$$

where n is the number of phenyl rings separating the two domain walls. E_{tot} increases due to the two ground states being non-degenerate, i.e. $\Delta E_0 \neq 0$. However theoretical work shows that in the presence of strong donors or acceptors charged confined soliton–antisoliton pairs may occur [8]. The magnetic susceptibility measurements in poly(*para*-phenylene) and polypyrrole show that the susceptibility is very small even when the polymers are doped in the highly conducting regime. The IR and photoemission measurements also show that midgap states are formed on doping these polymers. The spectroscopic measurements of polythiophene show two doping induced gap states situated symmetrically about the midgap. Because the soliton–antisoliton pairs in the c-PA must be confined, they combine quickly. This explains why photoconductivity is observed in the t-PA but not in the c-PA. However strong band edge absorption is observed in the c-PA.

Absorption and PL spectra of the thin layers of Ooct-OPV5 are shown in Fig. 3.5. The absorption peak has a much larger width than the PL peak. If the height of the long wavelength shoulder in the absorption spectrum is higher than the peak, the absorption and the PL will have mirror symmetry. The relative heights of these peaks can vary with experimental conditions. These results show that the behavior of the molecular solids is similar to the long chain conjugated conducting polymers. The general characteristics of both classes of materials are low mobility, wide absorption band and mirror symmetry between the absorption and the PL peaks. The band gap of Ooct-OPV5 determined from

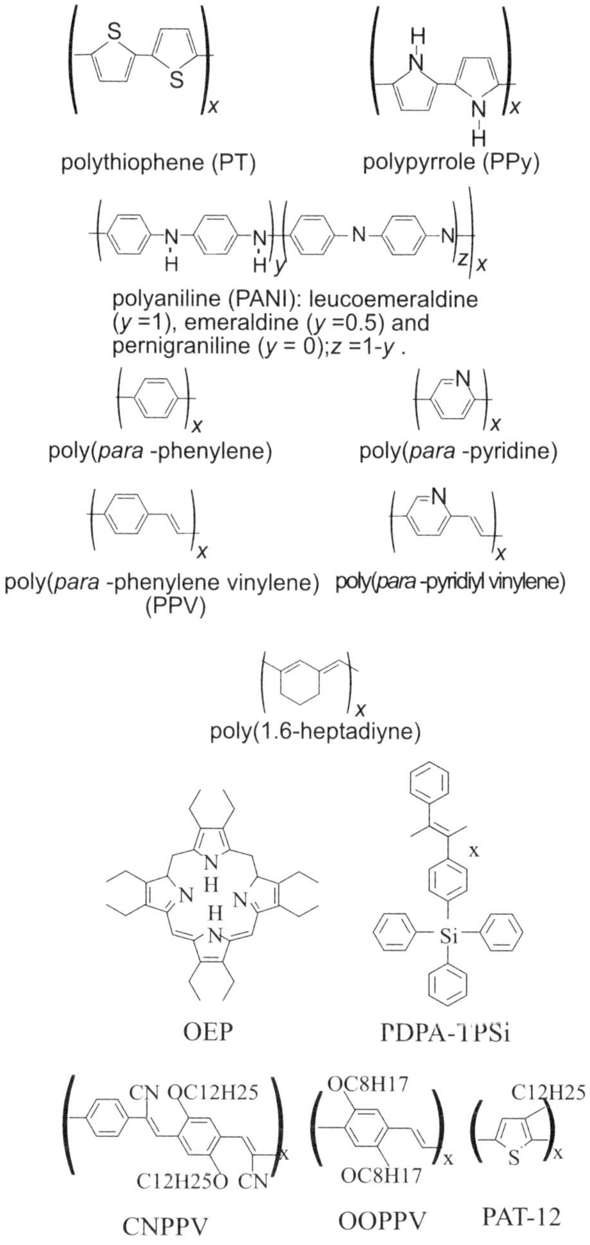

polythiophene (PT)

polypyrrole (PPy)

polyaniline (PANI): leucoemeraldine ($y=1$), emeraldine ($y=0.5$) and pernigraniline ($y=0$); $z=1-y$.

poly(*para*-phenylene)

poly(*para*-pyridine)

poly(*para*-phenylene vinylene) (PPV)

poly(*para*-pyridiyl vinylene)

poly(1.6-heptadiyne)

OEP

PDPA-TPSi

CNPPV

OOPPV

PAT-12

FIG. 3.2. Structure of several ring polymers with non-degenerate ground state.

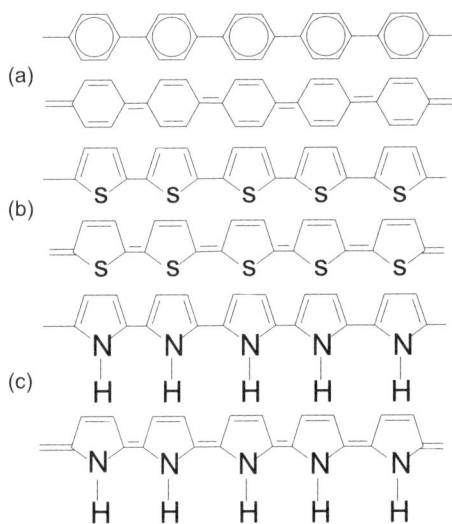

FIG. 3.3. Two nondegenerate ground states of (a) poly(*para*-phenylene), (b) poly(thiophene) and (c) polypyrrole [8].

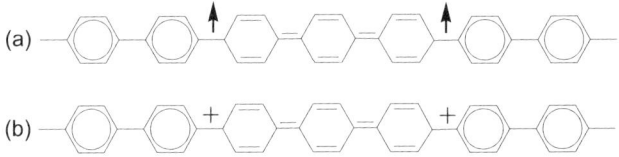

FIG. 3.4. (a) Neutral bipolaron (confined soliton–antisoliton pair) in poly(*para*-phenylene) and (b) bipolaron charged with charge $2|e|$ [8].

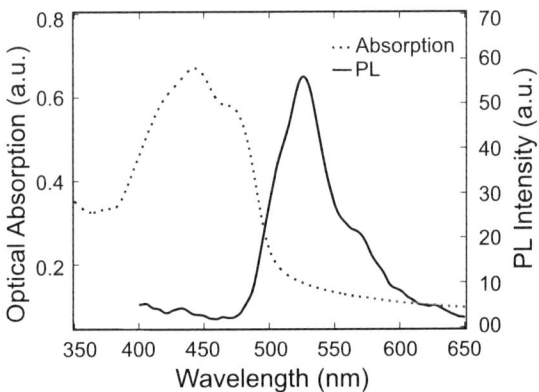

FIG. 3.5. Optical absorption and photoluminescence (PL) of a thin layer of Ooct-OPV5 [30].

TABLE 3.1
COMPARISON OF 300 K MATERIAL PROPERTIES OF SI, GAAS, 4HSIC AND GAN (THE DATA HAVE BEEN TAKEN FROM REF. [31] EXCEPT THOSE INDICATED IN THE TABLE FOOTNOTES)

Property	Si	GaAs	4H SiC	GaN
Breakdown field (10^5 V/cm)	2	4	30	30
Electron mobility (cm^2/V s)	1400	5000[a]	800	900[b]
Electron mobility (cm^2/V s)	–	8000[c]	–	1700[d]
Maximum velocity (10^7 cm/s)	1	2	2	3
Thermal conductivity (W/cm K)	1.5	0.5	4.9	1.3
Bandgap (eV)	1.1	1.42	2.45	3.4

Band gaps of conducting polymers cover the whole range from 1.4 to 3.4 eV

Mobilities in conducting polymers are extremely small and increase with applied electric field

[a] Ref. [32].
[b] $n = 10^{16}$ cm^{-3}.
[c] For an AlGaAs/GaAs heterostructure [32].
[d] For an AlGaN/GaN heterostructure [33].

TABLE 3.2
VALUES OF THE SMALLEST BANDGAP E_g OF THE CONDUCTING POLYMERS

Polymer	E_g (eV)	Ref.
trans-polyacetylene (t-PA)	1.4	[14]
cis-polyacetylene (c-PA)	2.0	[14]
Polypyrrole (PPy)	3.2	[14]
Polythiophene (PT)	1.6	[14]
Poly(para-phenylene) (PPP)	3.4	[14]
Poly(para-phenylene vinylene) (PPV)	3.0	[14]
Poly(para-phenylene vinylene) (PPV)	2.45	[34]
Polyaniline (PANI) leucoemeraldine form	3.0	[14]
MEH-PPV	2.1	[34]
BCH-PPV	2.2	[34]
O-PPV	2.13	[34]
PAT-6	2.2	[35]
PDPATPSi	2.7	[35]
OO-PPV	2.3	[35]
CN-PPV	2.3	[35]
OEP	1.8	[35]
C_{60}	1.8	[35]

the absorption edge is 2.4 eV. We have compiled parameters for materials used in blue and green light emitters in Table 3.1.

The minimum bandgaps derived from the optical measurements are given in Table 3.2.

3.2. The Dielectric Constant

Recent interest has been in the conductivity of more stable ring polymers (there are ordered and disordered regions, see Fig. 3.6). The microwave frequency measure-

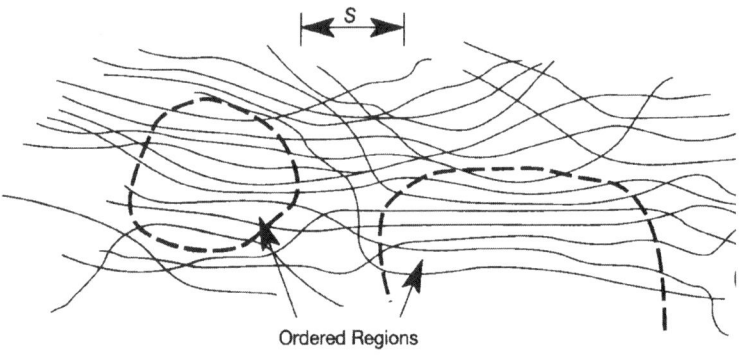

FIG. 3.6. Schematic diagram of the ordered and disordered regions. The ordered regions are 3–10 nm across. The width of the disordered region is shown by S [12].

ments of most heavily doped t-polyacetylene, PF_6 doped polypyrrole (PPY-PF_6), and d,l-camphorsulphonic acid doped polyaniline (PAN-CSA) give large negative values of the dielectric constant [12].

For PPY-PF_6 and PAN-CSA the microwave dielectric constant remains negative in the far IR even at 10^{-3} K, which shows that there are free carriers even at these low temperatures. These values of the dielectric constants give small values of plasma frequency which shows that only a small fraction of conduction electrons participate in plasma response. Scattering times come out to be 2 orders of magnitude larger than the values for alkali and noble metals. It is predicted that if technology improves, the conductivity of the doped polymers may become larger than that of metals.

3.3. Space Charge Limited Currents

3.3.1. Early Work of Mott. Poisson and Continuity Equations in a Trap-Free Insulator

In insulators practically there are no free electrons or holes and conductivity as given by Ohm's law is negligible. Mott [36] wanted to find out whether a current would flow in an insulator when a voltage is applied across it. His simple treatment given below shows that indeed a current will flow. However the injected carrier concentration decays rapidly in moving away from the injecting electrode. The injected carrier profile creates a strong electric field. The current that flows is determined by this space charge field and is known as the Space Charge Limited Current (SCLC). Almost all of the physics of modern organic devices is based on the SCLC first interpreted in trap-free solids by Mott in 1940 [36].

The band diagram of the metal–insulator contact is shown in Fig. 3.7(a) in thermal equilibrium and in Fig. 3.7(b) under an applied bias. The charge carrier density N_0 at the boundary (i.e. at the contact) in the insulator is given by [36],

$$N_0 \approx 10^{19} \exp[-(\phi - \chi)/kT]. \tag{3.3}$$

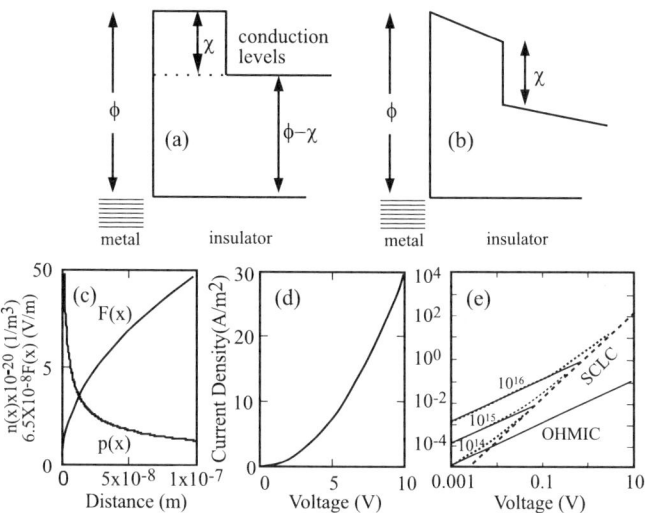

FIG. 3.7. (a) Metal–insulator contact in thermal equilibrium; (b) The contact under externally applied field; (c) electric field profile $F(x)$ and hole density $p(x)$ under an applied bias of 10 V; (d) J–V plot as given by Eq. (3.12). The thickness of the sample is 100 nm and a value of ϵ is 3; and (e) The effect of background doping concentrations (10^{14}, 10^{15} and 10^{16} cm^{-3}) on the current. The doping dominated regions at low voltages are ohmic shown by the straight lines.

Here ϕ is the work function of the metal and χ is the electron affinity of the insulator. Table 3.3 shows that N_0 is very sensitive to the values of $\phi - \chi$. In the organic polymer devices the injecting contact is made as nearly ohmic as possible and $\phi - \chi$ is small. In computation of the I–V relations $\phi - \chi$ is assumed to be zero. In this case the value of N_0 is very large and can be taken as infinity [37,38].

The current in the trap-free insulators is obtained by solving the following Poisson and continuity equations,

$$\frac{dF(x)}{dx} = \frac{q}{\epsilon \epsilon_0} p(x), \tag{3.4}$$

$$J = q \mu p(x) F(x). \tag{3.5}$$

Here q is the electronic charge, ϵ is the dielectric constant of semiconductor, and ϵ_0 is the permittivity of free space. The carrier density of holes[1] is denoted by p and the electric field by F. Mott showed that the diffusion component of the current is negligible and

TABLE 3.3
CALCULATED VALUES OF N_0 FOR THREE VALUES OF $\phi - \chi$

$\phi - \chi$ (eV)	0.1	0.5	1.0
N_0 (cm^{-3})	10^{17}	10^{10}	10^2

---

[1] We are considering a conducting polymer in which the current is dominated by the flow of holes and concentration of electrons is negligible. The treatment can be easily extended to the electron only samples.

need not be included in the continuity equation (3.5). The boundary condition is,

$$V = \int_0^d F(x)\,dx. \tag{3.6}$$

Here V is the applied voltage and d is the thickness of the sample. In this case the analytical solution can be easily obtained [36]:

$$F(x) = \sqrt{\frac{2J}{\epsilon\epsilon_0\mu}(x + x_0)}, \tag{3.7}$$

where x_0 is the constant of integration. The value of x_0 comes out to be,

$$x_0 = \frac{J\epsilon\epsilon_0}{2\mu N_0^2 q^2}. \tag{3.8}$$

The current–voltage relation is obtained by integrating the field $F(x)$ over the thickness d of the sample

$$V = \frac{2}{3}\sqrt{\frac{2J}{\mu\epsilon\epsilon_0}}\{(d+x_0)^{3/2} - x_0^{3/2}\}. \tag{3.9}$$

The carrier profile $p(x)$ is given by,

$$p(x) = \frac{\epsilon\epsilon_0}{2q}\sqrt{\frac{2J}{\mu\epsilon\epsilon_0}}\frac{1}{(x+x_0)}. \tag{3.10}$$

For $x_0 \gg d$, i.e. when V and J are large and/or N_0 is small, Eq. (3.9) reduces to Ohm's law,

$$J = \frac{q\mu N_0 V}{d}. \tag{3.11}$$

In practice this can happen in cases where $\phi - \chi \gg kT$. On the other hand for $\phi - \chi \ll kT$, Eq. (3.9) reduces to the well-known V^2 law,

$$J = \frac{9}{8}\epsilon\epsilon_0\mu\frac{V^2}{d^3}. \tag{3.12}$$

In the literature Eq. (3.12) is referred as the trap-free V^2-law, or Mott's V^2-law, or Child's law [37].

As mentioned earlier the condition $\phi - \chi \ll kT$ is generally assumed to hold [37–39, and references given therein]. This condition is equivalent to assuming that N_0 is very large and can be taken to be infinity. With this condition the mathematics is greatly simplified [37]. Now the electric field and carrier profiles are given by,

$$F(x) = \sqrt{\frac{2J}{\epsilon\epsilon_0\mu}x}, \tag{3.13}$$

and

$$p(x) = \frac{\epsilon\epsilon_0}{2q}\sqrt{\frac{2J}{\mu\epsilon\epsilon_0}\frac{1}{x}}. \qquad (3.14)$$

Calculated plots of Eqs. (3.13), (3.14), and (3.12), are shown in Figs. 3.7(c) and 3.7(d) respectively. The values of the parameters used in the calculations are $d = 100$ nm and $\epsilon = 3$. The applied voltage is 10 V in the calculations of field and carrier profiles. The above treatment is applicable to one carrier sample, i.e. the current is dominated either by the transport of electrons only or holes only.

3.3.2. EFFECT OF BACKGROUND DOPING

Let the background doping concentration be n_0 and let us assume that they are all ionized at the temperature of measurement. The concentration of existing holes due to background doping is also n_0. However their charge is compensated due to the ionized impurities. The total concentration of holes is p, the concentration of holes that contribute to the electric field is $p - n_0$. The Poisson equation (3.4) changes to,

$$\frac{dF(x)}{dx} = \frac{q}{\epsilon\epsilon_0}\bigl(p(x) - n_0\bigr). \qquad (3.15)$$

The coupled continuity equation (3.5) and Poisson equation (3.15) have been solved [40,37]. The solution combined with the boundary condition (3.6) yield the following relations,

$$w = u - \ln u, \qquad (3.16)$$

and

$$v = -\frac{1}{2}u^2 - u - \ln(1-u). \qquad (3.17)$$

The symbols u, v and w are functions of x. Here we are interested in the terminal characteristics and we give below expressions for these variables at $x = d$.

$$u_d = \frac{n_0}{p(d)} = \frac{qn_0\mu F(d)}{J}, \qquad (3.18)$$

$$v_d = \frac{q^3 n_0^3 \mu^2 V}{\epsilon\epsilon_0 J^2}, \qquad (3.19)$$

and

$$w_d = \frac{q^2 n_0^2 \mu d}{\epsilon\epsilon_0 J}. \qquad (3.20)$$

By replacing d with x in Eqs. (3.18), (3.19) and (3.20), the carrier density and field can be calculated as a function of x (see details and Fig. 1 of Ref. [40]). If we can eliminate u between Eqs. (3.16) and (3.17), a relation between current and voltage can be obtained. However it is not possible to do so. The J–V plots can be made by choosing several values of u and calculating v_d and w_d for each value. The plots made in this manner are shown in Fig. 3.7(e) for three background doping concentrations. The sample thickness is 130 nm and the values of other parameters are slightly different from those used in Figs. 3.7(c) and 3.7(d) [40]. At lower voltages, each curve is a (solid) straight line and agrees with the Ohm's law plot. As the voltage increases, the plots (closely spaced dot-curves) bend upwards. In the curved portion both ohmic and space charge limited currents contribute. As the voltage increases further the curves merge with the V^2 law plot (shown as a dashed line). Similar results are obtained with undoped narrow bandgap solids if thermal carrier density is large [37]. If the solid contains traps (discussed later), ohmic current is still obtained at lower voltages. At higher voltages appropriate SCLC law is obtained instead of the V^2 law.

The solid straight line corresponding to Ohm's law and the dashed line for the V^2 law meet at voltage V_{tr} given by,

$$V_{tr} = \frac{8qn_0 d^2}{9\epsilon\epsilon_0}. \tag{3.21}$$

The expression is useful for determining the background doping concentration n_0 by measuring V_{tr} experimentally. The expression for V_{tr} is different for a solid that has traps.

3.4. Polymers: The Solids with Traps

3.4.1. Poisson Equation with Trapped Charges

Both amorphous inorganic semiconductors and conducting organic materials have a large density of electron and hole traps. If the trap depth is large or temperature is low (i.e. if the trap levels are below the Fermi level), most of the injected carriers are trapped. A given applied voltage can support only a fixed quantity of total charge in the sample. Therefore the injected free carrier density is considerably reduced in the presence of traps. Both the trapped and the free charge carriers determine the space charge. The Poisson equation (3.4) changes to,

$$\frac{dF(x)}{dx} = \frac{q}{\epsilon\epsilon_0}\bigl(p(x) + p_t(x)\bigr), \tag{3.22}$$

where p_t is the trapped hole density. The transport equation (3.5) remains unchanged. Eqs. (3.5) and (3.22) are now solved to obtain the charge and field profiles and J–V relation for the conducting organic sample. We need to know the trap depth E_t to solve these equations. We will assume that the traps are uniformly distributed in space but may have a non-uniform distribution in the energy space. Four different distributions in the energy space have been considered in detail in the literature [37,38]: (1) all traps at a single energy level, (2) a Gaussian distribution of traps in the energy space, (3) exponentially distributed traps in the energy space and (4) uniformly distributed traps in the energy space. We consider the first three cases below. Kao and Hwang [38] have derived an expression for $J(V)$ in a polymer with uniformly distributed traps.

3.4.2. Single Level Traps

For single energy level traps the distribution can be written as,

$$h(E) = H_a \delta(E - E_t), \tag{3.23}$$

where H_a is the trap density and δ is the Dirac delta function. We first consider the shallow traps lying between the Fermi level and the valence band edge (we are considering conduction by holes). The concentration of holes in the traps is given by [38],

$$p_t = \int_{E_l}^{E_u} H_a \delta(E - E_t) \frac{1}{1 + \exp\left(\frac{E_F - E}{kT}\right)} dE. \tag{3.24}$$

Energies E_l and E_u are the lower and upper limits of the trapping level. This equation can be easily integrated [38,37] to give

$$p_t = \frac{H_a}{1 + \frac{H_a \theta_a}{p(x)}}, \tag{3.25}$$

where

$$\theta_a = \frac{N_v}{H_a} \exp(-E_t/kT), \tag{3.26}$$

and N_v is the effective density of states in the valence band. Using Eq. (3.25) for p_t and the boundary condition (3.6), the solution of the Poisson and transport equations gives [38],

$$J = \frac{9}{8} \theta_a \epsilon \epsilon_0 \mu \frac{V^2}{d^3}. \tag{3.27}$$

It can be easily shown that

$$\theta_a = \frac{p}{p + p_t}. \tag{3.28}$$

If there are no traps, $p_t = 0$, θ_a becomes equal to unity and Eq. (3.27) reduces to Eq. (3.12). Usually $p_t \gg p$ unless the applied voltage is very high. Under this condition θ_a becomes the ratio of free to trapped carriers.

At very low voltage the injected effective carrier density may be lower than the background thermal carrier density and the current is then ohmic. The voltage V_Ω at which SCLC begins to dominate is given by,

$$V_\Omega = \frac{8}{9} \frac{q p_0 d^2}{\theta_a \epsilon \epsilon_0}, \tag{3.29}$$

where p_0 is the impurity induced carrier density.

If the traps are deep, an analytical expression for J cannot be obtained. However numerical solutions can be obtained easily [38]. In this case most of the injected carriers remain trapped and the current by injection remains small until all the traps are filled. As the traps are nearly filled the SCLC begins to flow. It increases rapidly and as the trap-filled limit is reached, it follows the trap-free V^2 law. In this case trap-filled limit voltage V_{TFL} and V_Ω are equal. Recent work [41] on the approach to trap-filled limit will be discussed later.

3.4.3. GAUSSIANLY DISTRIBUTED TRAPS

In this case the distribution of traps in the energy space is given by,

$$h(E) = \frac{H_d}{(2\pi)^{1/2}\sigma_t} \exp\left[-\frac{(E-E_{tm})^2}{2\sigma_t^2}\right], \quad (3.30)$$

where $(2\pi)^{1/2}$ is the normalizing factor, E_{tm} is the trapping energy at the maximum trap density and σ_t is the standard deviation of the Gaussian distribution.

3.4.3.1. Shallow Gaussian Traps

In this case $E_{tm} < E_{F_p}$ and the ratio θ_d of free to trapped holes comes out to be [38],

$$\theta_d = \frac{p(x)}{p_t(x)} = \frac{N_v}{H_d} \exp\left[-\frac{E_{tm}}{kT} + \frac{1}{2}\left(\frac{\sigma_t}{kT}\right)^2\right]. \quad (3.31)$$

Using this relation, the Poisson and transport equations can be easily solved [38]. The expression for $J(V)$ is obtained by using the boundary condition (3.6),

$$J = \frac{9}{8}\mu\epsilon\epsilon_0\theta_d \frac{V^2}{d^3}. \quad (3.32)$$

Eq. (3.32) is similar to Eq. (3.27). The values of θ_a and θ_d are different but they are both constants independent of injection level. The voltage at which ohmic conduction changes to SCLC conduction is given by,

$$V_\Omega = \frac{8}{9} \frac{qp_0 d^2}{\theta_d \epsilon\epsilon_0}. \quad (3.33)$$

3.4.3.2. Deep Gaussian Traps

The trap depth for the deep traps is larger than the Fermi energy, i.e. $E_{tm} > E_{F_p}$. Using some approximations and assuming that $p_t \gg p$, the following expression for the electric field can be derived [37,38],

$$\frac{d[F(x)]^{(m+1)/m}}{dx} = \left(\frac{m+1}{m}\right)\frac{qH'_d}{\epsilon}\left(\frac{J}{q\mu_p N_v}\right)^{1/m}, \quad (3.34)$$

where

$$H'_d = \frac{1}{2}H_d \exp(E_{tm}/mkT), \quad (3.35)$$

and m is given by

$$m = \left(1 + \frac{2\pi\sigma_t^2}{16k^2T^2}\right)^{1/2}. \quad (3.36)$$

Solution of this equation along with the boundary condition (3.6) gives the J–V relation,

$$J = \frac{\mu_p N_v}{q^{m-1}}\left(\frac{2m+1}{m+1}\right)^{m+1}\left(\frac{m}{m+1}\frac{\epsilon\epsilon_0}{H'_d}\right)^m \frac{V^{m+1}}{d^{2m+1}}. \quad (3.37)$$

The expression for V_Ω for deep Gaussian traps is,

$$V_\Omega = \frac{qd^2 H'_d}{\epsilon\epsilon_0} \frac{m+1}{m}\left(\frac{m+1}{2m+1}\right)^{(m+1)/m}\left(\frac{p_0}{N_v}\right)^{1/m}. \tag{3.38}$$

3.5. Exponential Traps

3.5.1. Calculation of $J(V)$

Extensive work has been done on the exponentially distributed traps. We have devoted one separate section to discuss this work.

The concentration of the traps distributed exponentially in energy is given by,

$$h(E) = \frac{H_b}{kT_c}\exp\left(\frac{-E}{kT_c}\right). \tag{3.39}$$

H_b is the density of traps and T_c is the characteristic distribution constant. The concentration of trapped holes is given by,

$$p_t = \int_{E_{F_p}}^{\infty} \frac{H_b}{kT_c}\exp\left(\frac{-E}{kT_c}\right)\frac{1}{1+\exp[(E_{F_p}-E)/kT]}\,dE. \tag{3.40}$$

Denoting T_c/T by l, the above equation gives,

$$p_t = \alpha H_b \left(\frac{p}{N_v}\right)^{1/l}. \tag{3.41}$$

The factor $\alpha = (\pi/l)/\sin(\pi/l)$ (taken as unity in the calculations) arises because we have not used 0 K approximation (step function) for the Fermi occupancy function. Using this relation between free and trapped holes and assuming that the concentration of trapped holes is much larger than that of free holes, the coupled Poisson and transport equations can be solved. Using the boundary condition (3.6), the following analytical expression for $J(V)$ is obtained,

$$J = \frac{\mu_p N_v}{q^{l-1}}\left(\frac{2l+1}{l+1}\right)^{l+1}\left(\frac{l}{l+1}\frac{\epsilon\epsilon_0}{\alpha H_b}\right)^l \frac{V^{l+1}}{d^{2l+1}}. \tag{3.42}$$

Note that the solutions (3.37) and (3.42) are similar in form but the values of m and l are different.

3.6. Relaxation of the Approximation $p_t \gg p$

3.6.1. J–V Curves When $p_t \not\gg p$

Recently Jain et al. [41] have examined the validity of the approximation $p_t \gg p$ used in deriving Eq. (3.42). They calculated the hole profile in a polymer sample at two different applied voltages. The profiles are shown in Fig. 3.8. Fig. 3.8 shows that near

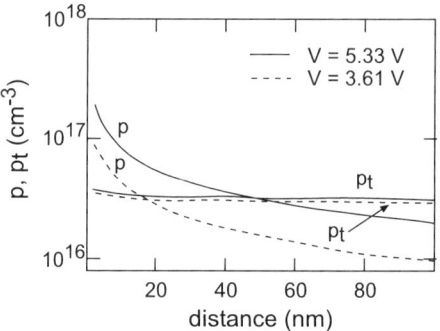

FIG. 3.8. Free hole density p and trapped hole density p_t as a function of distance from the contact at two applied voltages. Parameters used in the calculations are: $N_v = 10^{20}$ cm^{-3}, $H_b = 6 \times 10^{16}$ cm^{-3}, $T_c = 4000$ K, $\mu = 3 \times 10^{-5}$ cm^2 V^{-1} s^{-1}, $T = 300$ K, $\epsilon = 2$ and $d = 110$ nm.

the injecting end p is always larger than p_t because of the boundary condition that the injected carrier density is infinity at the contact. However at the lower applied voltage, p_t soon becomes larger than p. Since $p_t > p$ holds over most of the thickness of the sample, neglecting p in the Poisson equation becomes a valid approximation. The situation is quite different at the higher voltage. The free hole concentration p increases more rapidly than p_t and dominates over most of the sample thickness. Now the approximation $p_t > p$ breaks down. Without making the approximation $p_t > p$ the solution can be written in the form,

$$\int_0^d dx = \frac{\epsilon \epsilon_0}{q} \int_{F(x=0)}^{F(x=d)} \frac{dF}{\frac{J}{q\mu_p F} + \alpha H_b \left[\frac{J}{q\mu_p F N_v}\right]^{1/l}}. \tag{3.43}$$

The method to solve this type of equation numerically is described in Section 3.12. Numerical solutions of this equation show that at higher voltages the calculated J–V curves do not agree with Eq. (3.42).

3.6.2. TRAP-FILLED LIMIT

Early work on trap-filled limit is reviewed in Refs. [37,38]. It is believed that at some high voltage, designated as trap-filled voltage V_{TFL}, all the traps are filled and the current is given by the trap-free V^2 law. According to the existing literature [38], the V_{TFL} is determined by the voltage at the point at which the plots of Eq. (3.42) and of the V^2 law intersect and is given by [38],

$$V_{TFL,lit} = q \frac{d^2}{\epsilon \epsilon_0} \left(\frac{9}{8} \frac{(H_b)^l}{N_v} \left(\frac{l+1}{l} \right)^l \left(\frac{l+1}{2l+1} \right)^{l+1} \right)^{1/(l-1)}, \tag{3.44}$$

where additional subscript 'lit' emphasizes that the expression is taken from the existing literature. This approach to the trap-filling problem has a flaw. For the trap-free V^2 law to be valid, the electric field must be determined by the free injected charge carriers only. Therefore the V^2 law can be valid only if $p \gg p_t$. Also if $p_t \gg p$, Eq. (3.42) does not

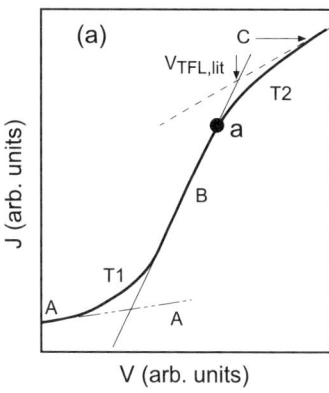
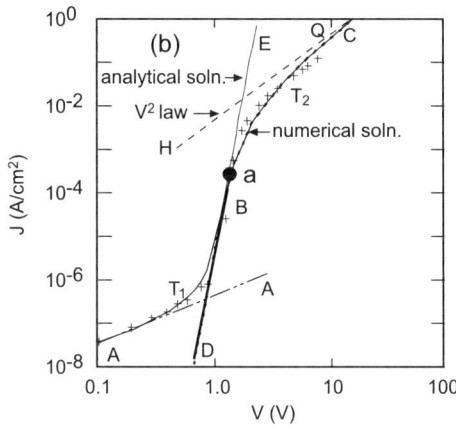

FIG. 3.9. (a) Schematic plot of typical J–V characteristic of a conducting organic material. The dashed line is the plot of the V^2 law. The dash-dot line is the plot of Ohm's law and straight line B is the SCLC. (b) Calculated and experimental plots of J–V characteristics of a PPV sample. DE is the plot of the analytical solution of Eq. (3.42). Curve DBC is obtained by numerical integration of Eq. (3.43). The figure is taken from Ref. [41].

remain valid up to the point of intersection. Therefore this procedure to determine V_{TFL} is not correct [41, see this reference for more details].

Schematic and calculated values of $J(V)$ curves are shown in Fig. 3.9. The lines marked A are plots of the ohmic law due to background doping of the sample [41,40]. The lines marked B show the SCLC 3.42 (3.42 is used to indicate that the calculations were made using Eq. (3.42)). The transition region T1 is obtained by the superposition of the region A and the SCLC region B. As the voltage increases the calculated current begins to deviate from the region B at the point marked 'a'. As the applied voltage approaches large values, the current approaches the trap-free V^2 law. There is a large transition region T2 between the region B and the V^2 law. The procedure to determine V_{TFL} given in the literature [38] is illustrated in Fig. 3.9(a). The straight line B, representing analytically calculated SCLC, is extrapolated to intersect the plot of the V^2 law. The point of intersection, marked $V_{\text{TFL,lit}}$, is assumed to be the value of the V_{TFL}. Fig. 3.9(b) shows that the analytical solution is not valid in the extrapolated region. At higher voltages beginning at point 'a', the free carrier density becomes large over significant part of the sample and now both the trapped and the free carriers determine the electric field. Numerical calculations show that the approximation $p_t \gg p$ used in deriving the equation for the region B does not remain valid beyond this point. Therefore the point marked $V_{\text{TFL,lit}}$ has no physical significance. The free carrier density keeps increasing with voltage and ultimately as $V \to \infty$, $p \gg p_t$ becomes valid and the influence of trapped carriers on the electric field becomes negligible. Therefore the V^2 law is obtained asymptotically only when V approaches infinity. The transport in the transition region T2 is also SCLC but both free and trapped carriers determine the electric field. The analytical solution (3.42) is incapable of predicting the transition region T2 and the voltage V_{TFL} at which the V^2 law becomes applicable. Theoretically V_{TFL} is infinitely large. For practical purposes we can define V_{TFL} as the voltage at which the difference between the calculated curve and the V^2 law is small and cannot be detected experimentally. 'Q' shows this point in Fig. 3.9(b).

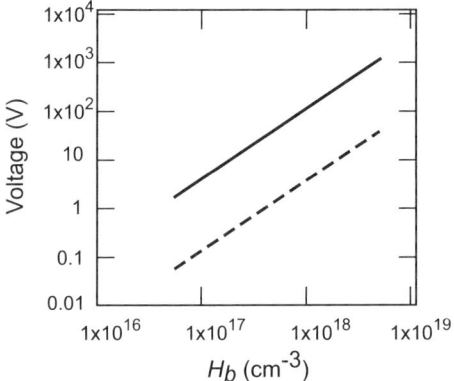

FIG. 3.10. The solid line represents the voltages of deviation from the V^2 law while the dashed line represents the voltages of deviation from Eq. (3.42). The other parameters used are: $T = 295$ K, $T_c = 1000$ K, $N_v = 2 \times 10^{19}$ cm^{-3}, $\alpha = 1$, $\mu = 1 \times 10^{-5}$ cm^2/V s, $\epsilon = 3$, and $d = 100$ nm.

Numerical calculations shown by the curve T1BT2C agree with recent experimental results [40] as shown in Fig. 3.9(b). The experimental data was taken on the diodes based on the five-ring PPV-type oligomer 2-methoxy-5-(2'-ethylhexyloxy)-1,4-bis((4',4"-bisstyryl)styrylbenzene (abbreviated as MEH-OPV5). The MEH-OPV5 films were sublimed in UHV (10^{-9} Torr) on a cooled ITO coated glass plate. The substrate was pre-patterned and the source temperature was 210 °C. Substrate-cleaning steps were done in a clean room. The top electrode was Al evaporated through a shadow mask. The active device areas ranged between 0.02 and 0.64 cm^2. The processing steps were performed under nitrogen atmosphere. The measured J–V values of the ITO/MEH-OPV5/Al diode (active area = 0.32 cm^2) are shown by the + symbols in Fig. 3.9(b). The experimental data, shown by the symbols, deviates from the plot of analytical solution (3.42) at both low and high voltages. The low voltage deviation occurs due to the presence of background impurities as discussed earlier [40]. The dash-double dot line shows the calculated ohmic current. The data merge with the straight-line plot of Eq. (3.42) when the space charge limited current dominates. The agreement of the experimental data with the numerical solution (3.43) over the whole voltage range, including the region T2, is excellent.

The dashed curve for a given set of parameters in Fig. 3.10 shows the calculated voltages at which the current deviates from the conventional theory given by Eq. (3.42) and approaches to V^2 law (within 10% error) as a function of total trap density H_b.

3.7. Effect of Finite (Non-Zero) Schottky Barrier

3.7.1. Importance of Finite Barriers

The theory of transport in insulators (including conducting organics) is about fifty years old [36,37, and references given therein]. This theory assumes zero barriers for injection from metal into the polymer, i.e. the contact is ohmic. The theory based on ohmic contacts has been used extensively in the literature for transport in organics [42,43,41]. For most cathode/anode materials and organics, the Schottky barrier ϕ_B is not zero. Since no theory of bulk transport in organic conducting materials exists for this case the experimental data is always compared with the theory for zero Schottky barriers. We have developed a theory of bulk transport for finite (non-zero) Schottky barriers and have calculated for the first time the charge carrier density distribution, the electric field distribution and the J–V characteristics for these cases [44]. The results are strikingly different from those obtained for zero barriers (i.e. ohmic contact).

Several authors have observed that log J versus log V plots are straight lines. This result is consistent with the conventional power law (3.42). It is not clear how this result is modified for non-zero Schottky barriers. The J–V characteristics in our theory depend on the value of the constant of integration C. The Schottky barriers ϕ_B define $P(0)$, and C depends on $P(0)$, where $P(0)$ is the injected charge carrier density at the contact.

3.7.2. Theory

Assuming exponential trap distribution, combining continuity and Poisson equations and integrating the resultant equation, we obtain,

$$F(x) = \left(\frac{qH_b}{\epsilon\epsilon_0}\frac{(l+1)}{l}(x+C)\right)^{\frac{l}{l+1}}\left(\frac{J}{q\mu N_v}\right)^{\frac{1}{l+1}}, \qquad (3.45)$$

where C is the constant of integration. As mentioned before in earlier published work the Schottky barrier was taken to be zero (i.e. $C = 0$) [38]. Kumar et al. [44] show that in practical cases neglecting C can cause large errors. Taking $\alpha = 1$ and using the boundary condition (3.6) we get

$$J = \frac{\mu_p N_v}{q^{l-1}}\left(\frac{2l+1}{l+1}\right)^{l+1}\left(\frac{l}{l+1}\frac{\epsilon\epsilon_0}{H_b}\right)^l \frac{V^{l+1}}{[(d+C)^{\frac{2l+1}{l+1}} - C^{\frac{2l+1}{l+1}}]^{l+1}}. \qquad (3.46)$$

For a given J (or applied voltage V) C is independent of x and $F(x)$. C is different for different values of J. We now determine the constant C in terms of material parameters and J. Since at the injecting contact $x = 0$, the field $F(0)$ can be determined in terms of J from the continuity equation and $P(x) = P(0)$, the value of C is given by

$$C = J\left[\frac{1}{q\mu P(0)}\right]^{\frac{l+1}{l}}\left(\frac{\epsilon\epsilon_0}{qH_b}\right)\left(\frac{l}{l+1}\right)(q\mu N_v)^{\frac{1}{l}}. \qquad (3.47)$$

Above equation shows that for infinitely large $P(0)$, C becomes zero. C decreases with increase in H_b and l and increases with J (or applied voltage). The effect of temperature comes through $l = T_c/T$. Numerical calculations show that as l increases, C decreases. At low temperatures l is larger and C becomes smaller. The effect of C becomes more pronounced at higher temperatures. The value of $P(0)$ is determined by the Schottky barrier ϕ_B [36]. We assume that the injected charge density at the contact remains constant at its thermal equilibrium value [45, see p. 258] when a current flows through the sample. If $\phi_B = 0$, the injected hole density $P(0) = 10^{20} \approx \infty$ and $C = 0$, (3.46) reduces to Eq. (3.42) (for $\alpha = 1$) discussed earlier. As mentioned earlier as C increases to > 0, the J–V characteristics deviate considerably from Eq. (3.42).

In Eq. (3.46) two cases of interest arise. The first case is when C is smaller than d. When C is so small that it can be neglected in comparison with d in Eq. (3.46), the J–V characteristics are given by Eq. (3.42). If C cannot be neglected compared to d, the J–V curves begin to deviate from the power law. The second case when $C > d$, is of practical interest. Expanding the denominator in Eq. (3.46) by binomial expansion in this case we obtain

$$J = q\mu P(0)\frac{V}{d}\left(1 + \frac{l}{(2l+1)}\frac{d}{C}\right)^{-1}. \tag{3.48}$$

For $C \gg d$ the second term in Eq. (3.48) can be neglected and we obtain

$$J = q\mu P(0)\frac{V}{d}. \tag{3.49}$$

This result is important since it shows that for finite (non-zero) Schottky barrier, i.e. $P(0) < \infty$, and for large value of C, the current changes from the space charge limited current to the ohmic current. This result is similar to that given by Mott [36] for a trap-free insulator. Just before the ohmic region there is a transition region between the conventional power law and the ohmic region.

3.7.3. Results and Discussion

We now calculate the J–V characteristics of an organic diode using the equations of the previous section. The values of the constant of integration C as a function of $P(0)$ for different values of l are shown in Fig. 3.11(a) and for different values of J in Fig. 3.11(b). The constant C decreases rapidly as the injected charge density increases for all values of J and l. As ϕ_B tends to zero, $P(0)$ tends to infinity and the constant C becomes zero. We have already mentioned that now the J–V characteristics of an organic Schottky

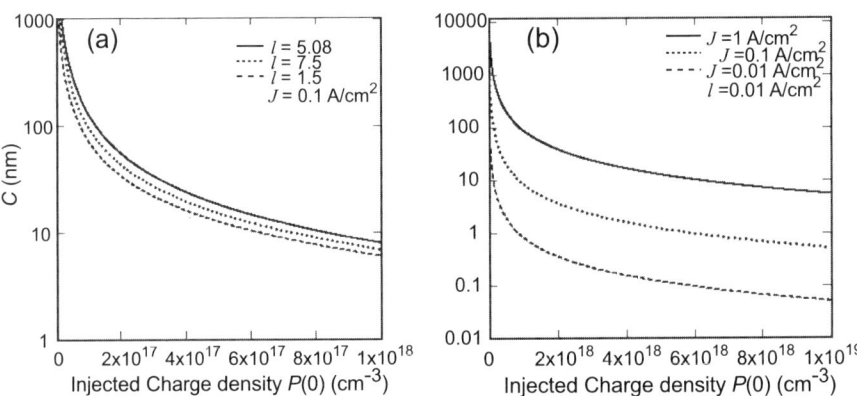

FIG. 3.11. (a) The effect of $P(0)$ on the constant of integration C for different values of $l = T_c/T$. The calculations have been made for $J = 0.1$ A/cm^2. (b) The effect of $P(0)$ on the constant of integration C for different values of J. The calculations have been made for $l = 5.08$. As $P(0)$ tends to infinity, C tends to zero and Eq. (3.42) becomes valid. The values of the other parameters used in the calculations are: $H_b = 2 \times 10^{17}$ cm^{-3}, $N_v = 2 \times 10^{19}$ cm^{-3}, $\epsilon = 3$, $\epsilon_0 = 8.85 \times 10^{-14}$ F/cm, $\mu = 1 \times 10^{-5}$ cm^2/V s, $d = 100$ nm, $T_c = 1500$ K, $T = 295$ K and $q = 1.6 \times 10^{-19}$ C [44].

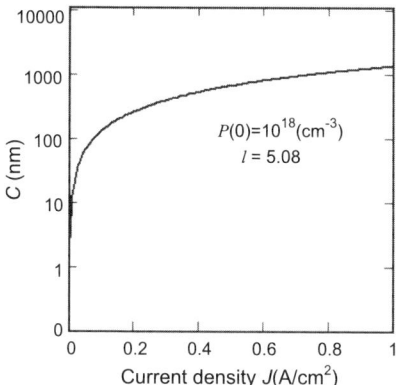

FIG. 3.12. The effect of current density on the constant of integration C at injected charge density $P(0) = 10^{18}$ cm^{-3}. Since voltage increases with J, C increases with applied voltage also. The values of the other parameters are the same as given in Fig. 3.11 [44].

diode obeys the old conventional theory given by Eq. (3.42). Fig. 3.12 shows that for a given $P(0)$ the constant C increases with current density. The field distribution and charge carrier distribution along the direction of current flow are important quantities. Their calculated values are plotted in Figs. 3.13 and 3.14.

As $P(0)$ decreases (or C increases) both the field and charge carrier densities tend to become uniform. When the field and charge densities become uniform one would expect the ohmic relationship between current and voltage. This is consistent with the result derived earlier in Eq. (3.49). The effect of $P(0)$ (or ϕ_B because $P(0)$ varies as $\exp(-\phi_B/kT)$) is shown by the calculated J–V curves in Fig. 3.15. Curves A represent the J–V characteristics and correspond to the conventional theory equation (3.42), while curves B are the J–V curves corresponding to the new theory equation (3.46). The curves B depend on the constant of integration C. We have already shown that C decreases as

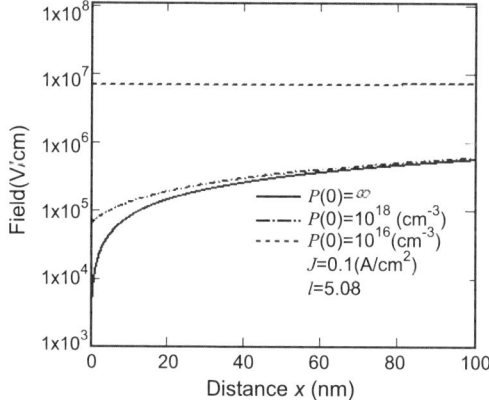

FIG. 3.13. Electric field as a function of distance x from the injecting contact for different values of $P(0)$. The values of the other parameters used in the calculation are the same as given in Fig. 3.11. For small value of $P(0)$, $P(0) = 10^{16}$ cm^{-3} ($\phi_B = 0.24$ eV) the field becomes practically independent of x.

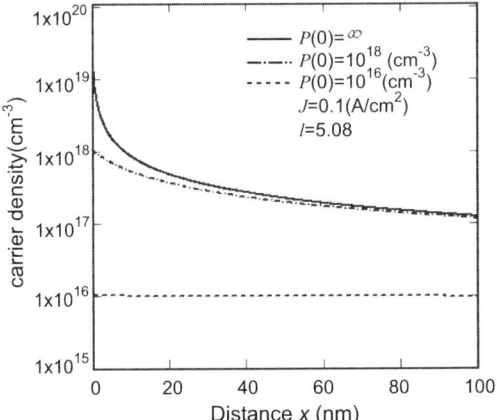

FIG. 3.14. Carrier density as a function of distance x from the injecting contact for different values of $P(0)$. The values of the other parameters used in this calculation are the same as given in Fig. 3.11. Note that for small value of $P(0)$, $P(0) = 10^{16}$ cm^{-3} ($\phi_B = 0.24$ eV), the carrier density becomes practically independent of x [44].

$P(0)$ increases but C increases as J or V increases. It is therefore easy to understand the behavior of curves in Fig. 3.15. For small C (large $P(0)$ or/and small J) the curves using present theory become practically identical with the conventional theory equation (3.42). As C increases (i.e. for smaller $P(0)$ or/and larger J) the current deviates from the plot of Eq. (3.42). For sufficiently large value of C the curves B become straight lines which correspond to Ohm's law. The ohmic and transition region in curves B are well described by the new theory.

The voltages at which the J–V curve deviates from the power law V^{l+1} and at which it approaches Ohm's law are of general interest. Calculations have been made for a typical case, a 10% difference is taken as the experimental tolerance. The results are shown in Fig. 3.16. The voltages of deviation from the power law (dashed curve) and that of approach to Ohm's law (solid curve) are shown for different values of the injection barrier. It is seen that these voltages are large for small value of the barrier and decrease rapidly as the barrier increases.

3.7.4. COMPARISON WITH EXPERIMENT

Typical experimental data measured at 240 K is shown by symbols in Fig. 3.17. The dash-dot curve is the plot of Ohm's law at low voltages. The experimental data agrees with Ohm's law upto the point shown by B. This ohmic behavior of the J–V characteristics at low voltages is due to existing background doping or thermally generated carriers. The ohmic region is followed by transition to the space charge limited current (SCLC) and the SCLC follows the V^{l+1} law, characteristics of exponential trap distribution. The dotted curve ADF is the plot of new theory equation (3.46). Dashed line ADC is the plot of old theory given by Eq. (3.42). The two theories agree upto the point D and the experimental data is in good agreement with both the theories between the points E and D. As voltage increases beyond point D, constant C becomes large and the new theory deviates from the

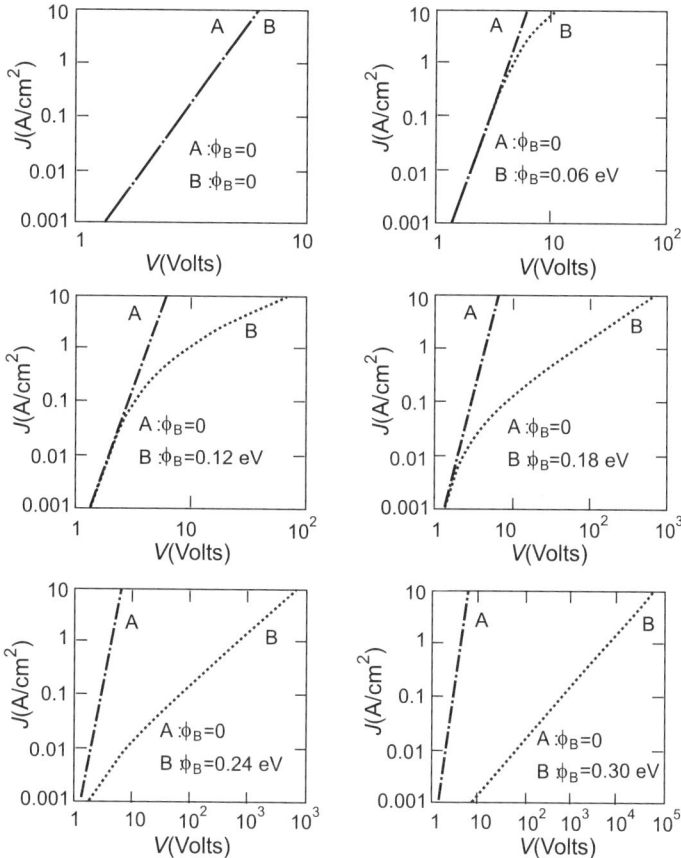

FIG. 3.15. J–V characteristics of Schottky diodes with $\phi_B = 0$ and for finite (non-zero) values of ϕ_B are plotted. Curves A correspond to conventional equation (3.42) and curves B correspond to Eq. (3.46) for different values of ϕ_B. For a non-zero Schottky barrier curves B deviate from curves A as applied voltages increases. Curves B become straight lines at higher voltages where C is large. These straight lines correspond to Ohm's law as predicted by Eq. (3.49). The values of the parameters in these calculations are the same as given in Fig. 3.11 [44].

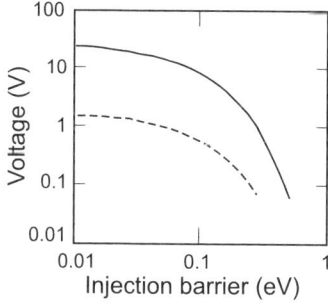

FIG. 3.16. The voltage at which the J–V curve deviates from the power law and that at which it approaches Ohm's law, within 10% [44].

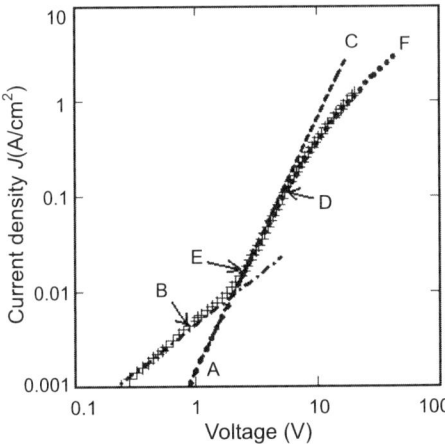

FIG. 3.17. J–V characteristics of the ITO/PEDOT:PSS/MEH-PPV/Au diode at 240 K. The thickness of the active layer is 120 nm. The symbols represent the experimental data. The dash-dot line represents the ohmic region due to thermally generated and background carriers. The dashed line represents the calculated values using the conventional equation (3.42) with a zero Schottky barrier, while the dotted curve represents the calculated values using Eq. (3.46) with a Schottky barrier = 0.1 eV. At lower voltages below the point D, the plot of Eqs. (3.46) and (3.42) are practically identical. The values of the parameters used in this representation are: $N_0 = 10^{19}$ cm^{-3}, $T_c = 400$ K, $\epsilon = 3$, $\epsilon_0 = 8.85 \times 10^{-14}$ F/cm, $\mu = 7 \times 10^{-5}$ cm^2/V s, $N_v = 2 \times 10^{19}$ cm^{-3} and $H_b = 4.5 \times 10^{18}$ cm^{-3} [44].

old theory. The experimental data agrees with the new theory. As the voltage increases further the value of $l + 1$ (in $J \sim V^{l+1}$) decreases and at sufficiently high voltages it tends to become 1 corresponding to Ohm's law. The parameter $P(0)$ comes out to be 9×10^{16} cm^{-3}, which corresponds to a Schottky barrier of 0.1 eV. Taking the HOMO of MEH-PPV at 5.3 eV and work function of PEDOT:PSS as 5.2 eV the expected barrier is 0.1 eV, which is in excellent agreement with that derived with the theory.

3.8. Combined Effect

It is of interest to examine the case when both effects are taken into account, i.e. the injection barrier is not zero and also $p_t \not\gg p$. Such cases can arise in practice only if H_b is small. To a non-zero Schottky barrier the value of $P(0)$ is reduced. Since p is always smaller than $P(0)$, a value of p is also reduced. In general this results in $p_t > p$ and the power law remains valid and Ohm's law is approached at high voltages. The numerically calculated values of the J–V curves for such a case are shown in Fig. 3.18.

3.9. Temperature Effects

3.9.1. Temperature Effects in PPV-Based Polymers

Campbell and co-workers [46] have measured the J–V curves of PPV-based polymers at different temperatures in the range 30–270 K. They were able to fit the trapping model,

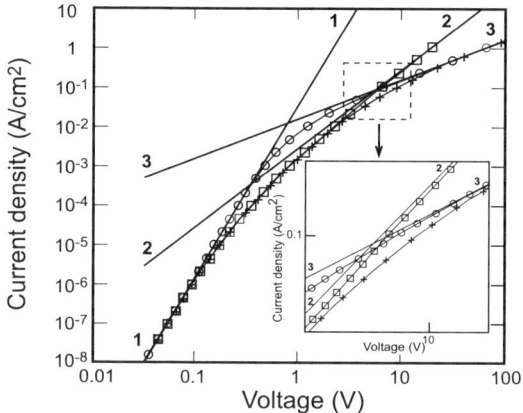

FIG. 3.18. Calculated J–V characteristics of an organic diode with exponentially distributed traps in the energy space. The solid lines 1, 2 and 3 correspond to the conventional theory equation (3.42), the V^2 law equation (3.12) and the Ohm's law equation (3.46) respectively. Open circles represent the J–V characteristic for finite (non-zero) injection barrier with $p < p_t$, while open squares represent the J–V characteristic with zero injection barrier but $p \not< p_t$. The plus symbols represent the J–V characteristic when both the cases exist, injection barrier is finite (non-zero) and $p \not< p_t$. The values of the parameters used are: $T = 295$ K, $T_c = 1000$ K, $N_v = 2 \times 10^{19}$ cm^{-3}, $H_b = 1 \times 10^{17}$ cm^{-3}, $\mu = 1 \times 10^{-5}$ cm^2/V s, $\epsilon = 3$, $\epsilon_0 = 8.85 \times 10^{-14}$ F/cm, $d = 100$ nm and the injection barrier $\phi_B = 0.11$ eV. Inset shows the magnified image of the marked portion.

Eq. (3.42), to their data but the values of the parameters used were somewhat different for different temperatures. A closer examination of the data (shown in Fig. 3.19(a)) shows that the trapping model does not explain the results well. For example theoretically by changing the temperature from 270 to 30 K the value of $l + 1$ should change by a factor of 9. Experimentally the change was only by a factor of 2. We have tried to fit the trapping model with their data [42]. The values of the parameters used in the calculations are: $T_c = 1920$ K, $H_b = 9 \times 10^{17}$ cm^{-3}, $N_v = 3 \times 10^{19}$ cm^{-3}, $\epsilon = 2$ and

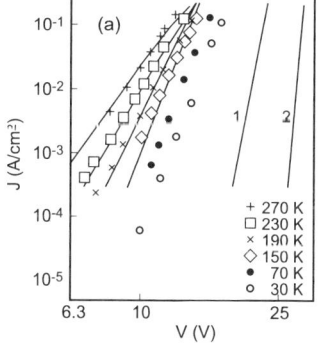
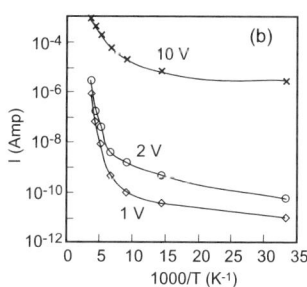

FIG. 3.19. (a) J–V characteristics of a sample of MEH-PPV with a thickness of 94 nm. The symbols represent the experimental data from Campbell et al. [46] at different temperatures. The solid straight lines are the calculated values using the trapping model in the same order of temperature from 270 to 30 K. (b) The same data re-plotted: values of I plotted as a function of $1/T$ for $V = 10, 2$ and 1 V. The figure is taken from [42].

$\mu = 9 \times 10^{-5}$ cm^2 V^{-1} s^{-1}. At 270 K the fit of the trapping theory is satisfactory as shown in the figure. The theory did not agree with the experimental data at lower temperatures if the same values of the parameters are used. By increasing the mobility to 1.7×10^{-4} and 6×10^{-4} cm^2/V s the data could be made to agree with the theory at 230 and 190 K as shown in the figure. Below 190 K the discrepancy between the calculated and experimental curves increased very rapidly. The discrepancy at the two lowest temperatures shown by the data (symbols) and calculated results (solid lines 1 and 2) is very large. Now changes in the other parameters are necessary to bring agreement between theory and experimental data. Since the measurements were made on the same sample it is not reasonable to expect that T_c and H_b are different at different temperatures.

Campbell et al. [46] also compared their data with the mobility model. The mobility model showed a very steep rise in the current with voltage at low temperature, which is inconsistent with their data. More recently Berleb et al. [47] and Lupton et al. [48] have also measured the J–V curves of conducting organic semiconductors at different temperatures. Similar discrepancies between theory and their data are found.

Arrhenius plots of current versus $1/T$ in Fig. 3.19(b) show that each plot consists of approximately two straight lines: one at high temperatures with higher activation energy and the other at lower temperatures with lower activation energy. These results suggest that the trapping model determines the $J(V)$ in the high temperature region. At the high temperatures, the traps are ionized and holes can reach the extended states [42] where they drift as free holes. At the low temperatures, the thermal energy is not sufficient and holes remain trapped. The trapping model as formulated above is not applicable under these conditions. The holes however can tunnel from one trap to another at low temperatures, which suggests that the mobility model should be preferred now.

3.9.1.1. Analytical Derivation of the Arrhenius Behavior

It is not clear from Eq. (3.42) that the J–$1/T$ plots should be linear and Arrhenius like as observed in the high temperature regime. Fig. 3.20(a) shows theoretically calculated plots of log J–$1/T$. The calculations were made by numerically integrating Eq. (3.43), which is free from the approximations used in deriving Eq. (3.42). Fig. 3.20(a) shows that indeed the log J–$1/T$ plots are linear. The activation energy determined from the slope depends on the value of T_c. The effective activation energy E_{eff} determined from the slopes varies linearly with the trap energy $E_t = kT_c$ as shown in the inset of the figure. Comparison of the mobility model with the low temperature data is shown in Fig. 3.20(b). The agreement between theory and experiment is excellent. However there is one weakness in these results. In theoretical calculations different values of H_b were used at different temperatures in the mobility model. Since the tunneling probability depends on both the depth of the trap and temperature, it is possible that different number of traps is active at different temperatures.

The assumptions and approximations used in deriving Eq. (3.42) break down at low temperatures or at high-applied voltages. At room temperature and at low applied voltage the values of $J(V)$ obtained using this equation agree closely with the numerical solutions. Therefore Eq. (3.42) should also yield Arrhenius straight lines if log I values are plotted as a function of $1/T$. Recent work has shown that this is indeed the case [49].

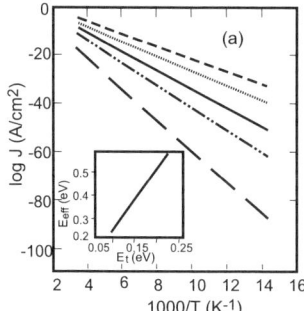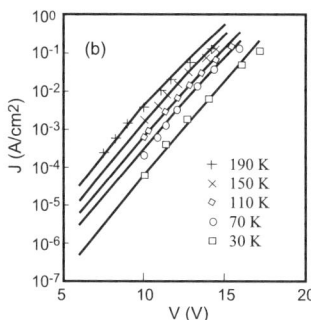

FIG. 3.20. (a) Calculated current density as a function of inverse temperature for $V = 10$ V for a conducting polymer sample with exponentially distributed traps. The values of the trap distribution parameter T_c are: $T_c = 2500$ K (long dash line), $T_c = 1800$ K (dash-dot line), $T_c = 1500$ K (solid line), $T_c = 1200$ K (dotted line) and $T_c = 1000$ K (small dash line). The values of the other parameters are: $H_b = 3 \times 10^{18}$ cm^{-3}, $N_v = 3 \times 10^{20}$ cm^{-3}, $\epsilon = 2$ and $\mu_p = 5 \times 10^{-6}$ cm^2 V^{-1} s^{-1}. The inset shows the calculated effective activation energy, E_{eff}, as a function of the characteristic trap distribution energy $E_t = kT_c$. (b) J–V characteristics of a sample of MEH-PPV with a thickness of 94 nm on a log-linear scale. The symbols represent the data of Campbell et al. [46]. The lines represent calculated values using the mobility model. The figure is taken from [42].

After some algebraic manipulation Eq. (3.42) can be written as,

$$J = \left(\frac{\mu N_v q V}{d}\right) f(l) \exp\left[-\frac{E_t}{kT} \ln\left(\frac{q H_b d^2}{2\epsilon \epsilon_0 V}\right)\right], \quad (3.50)$$

where

$$f(l) = \left(\frac{2l+1}{l+1}\right)^{l+1} \left(\frac{l}{l+1}\right)^l \frac{1}{2^l}. \quad (3.51)$$

It can be easily seen that the function $f(l)$ becomes practically constant (i.e. independent of l) at 0.5 for $l > 2$. In practice l is always more than 2. Therefore Eq. (3.50) can be written as,

$$J = \left(\frac{\mu N_v q V}{2d}\right) \exp\left[-\frac{E_t}{kT} \ln\left(\frac{q H_b d^2}{2\epsilon \epsilon_0 V}\right)\right], \quad (3.52)$$

to a good approximation. This equation predicts that analytically calculated plots of $\ln J$–$1/T$ are Arrhenius straight lines. However the activation energy determined from the slopes of these lines are not the activation energy E_t but the energy multiplied by a factor γ,

$$\gamma = \ln \frac{q H_b d^2}{2\epsilon \epsilon_0 V}, \quad (3.53)$$

$$E_{t(\text{eff})} = \gamma E_t. \quad (3.54)$$

The activation energy observed experimentally is the effective activation energy and is a function of applied voltage V and trap density H_b. Several interesting features predicted by this equation are given below.

1. For small applied voltage γ is positive and the slope of the $\ln J$–$1/T$ plots is negative. For sufficiently high applied voltage V, γ becomes negative and the slopes, positive. The change-over takes place at applied voltage $V = V_c$ given by,

$$V_c = \frac{qH_b d^2}{2\epsilon\epsilon_0}. \tag{3.55}$$

At $V = V_c$ the effective activation energy is zero.
2. The $\ln J$ versus $\ln V$ plots measured at different temperatures also show interesting features. At $V = V_c$ the curves are independent of temperature. Therefore the curves plotted at different temperatures cross each other at one point at $V = V_c$.
3. For determining H_b, it is sufficient to measure $J(V)$ at few temperatures, extrapolate them until they meet and determine V_c. H_b can now be determined using Eq. (3.55). The trap density can also be determined from the slopes of $\ln J$–$1/T$ plots using Eqs. (3.54) and (3.53). Trap densities determined by these simple direct methods agree with those derived by complicated fitting procedures [49].

We have discussed the band model of SCLC. In this model the charge carriers move by drift in the LUMO (or conduction band). In some cases the observed current has been attributed to tunneling through the Schottky barrier. Parker [50] has studied the MEH-PPV and suggested that the J–V relationship of this material is determined by tunneling of the charge carriers through the interface barriers. One other model, known as the mobility model has also been used to interpret the charge transport in the polymers. Blom and co-workers [51] studied the J–V characteristics of some PPV derivative and suggested that conduction is space charge limited, and it can be interpreted by the mobility model. The mobility model is discussed in detail in Section 3.10. Campbell and co-workers [46,52] studied a PPV derivative and concluded that in case of single carrier injection, the tunneling theory shows very poor agreement. They found that the mobility model agreed with experimental data only in very limited range of J and temperature. Power law ($J \sim V^{l+1}$) based on the band model agreed with experiment over a considerable range of applied voltages [46]. Kumar et al. [44] found that the results of their hole conduction measurements in MEH-PPV at low and high applied voltages and at different temperatures agree with the band model. At very high voltages, the current deviates from this law.

3.9.2. Recent Work

The J–V curves of MEH-PPV in hole only devices, i.e. in ITO/PEDOT:PSS/MEH-PPV/Au structure, have been studied recently. The J–V characteristics in the temperature range from 300 to 98 K are shown in Fig. 3.21. The experimental J–V characteristics and their fitting with the calculated J–V curves are shown in Fig. 3.21. The symbols represent the experimental data while the solid lines represent the calculated J–V curves. At low voltages the current shows ohmic behavior which is due to thermally generated or background carriers. At some intermediate voltages the current increases rapidly. As the voltage increases further, the contribution of free carriers does not remain negligible, the current deviates from Eq. (3.42). Our attempts to obtain the V^2 law near room temperatures were not successful. In fact it can be seen from Fig. 3.21 that the slopes

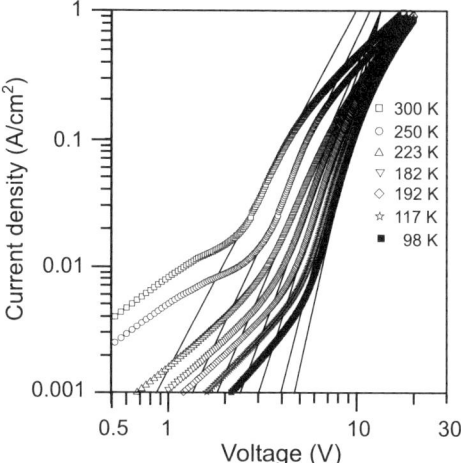

FIG. 3.21. $J-V$ characteristics of the conducting organic polymer diode ITO/PEDOT:PSS/MEH-PPV/Au on the log–log scale at different temperatures, where the thickness of MEH-PPV is 65 nm. The symbols represent the experimental data, while the solid lines are the calculated values using Eq. (3.42) at the corresponding temperatures [53].

of higher temperature $J-V$ characteristics are smaller than 2 at higher voltages. Since the slopes are less than 2 it is not possible to obtain the V^2 law whatever large the voltage is. This is surprising result since the work of Jain et al. [41] shows that the V^2 law should be obeyed at high voltages. This shows that there is some other physical factor, which comes into play and has not been taken into account in the theory of Jain et al. The theory is based on the assumption that the carrier density at the injecting contact is infinitely large. In fact the Schottky barrier at ITO-PEDOT:PSS/MEH-PPV interface is about 0.1 eV, which shows that the carrier density at the injecting contact is $P(0) = 10^{19} \exp(-0.1/kT) = 2.1 \times 10^{17}$ cm^{-3}. Numerical calculations show that the $J-V$ curves for $P(0) = 2.1 \times 10^{17}$ cm^{-3} are considerably different from those for $P(0) = 10^{19}$ cm^{-3}.

At low temperatures charge carriers do not have sufficient energy to go into the extended states and the traps tend to fill completely. As soon as the traps are filled p tends to become much more than p_t and currents follow the V^2 law. For the clarity and complete explanation of our low temperature experimental data the $J-V$ characteristic at 98 K has been shown separately in Fig. 3.22. The high voltage region in Fig. 3.22 corresponds to the V^2 law. The mobility has been calculated from this region to be 0.2×10^{-5} cm^2 V^{-1} s^{-1}, which is comparable to the earlier reported values. The values of other parameters used are: $H_b = 1 \times 10^{18}$ cm^{-3}, $T_c = 550$ K, $N_v = 2.7 \times 10^{18}$ cm^{-3}, $\epsilon = 3$ and $\epsilon_0 = 8.85 \times 10^{-14}$ F/cm. The experimental data is in good agreement from 98 to 223 K with the theory equation (3.42) for the same values of the parameters used. But the experimental data at 250 and 300 K is in good agreement with the theory if the mobility is slightly increased to 0.31×10^{-5} cm^2/V s and 0.47×10^{-5} cm^2/V s respectively. This increment in the mobility with temperature is quite low and mobility can be assumed to be independent of the temperature in the whole temperature range.

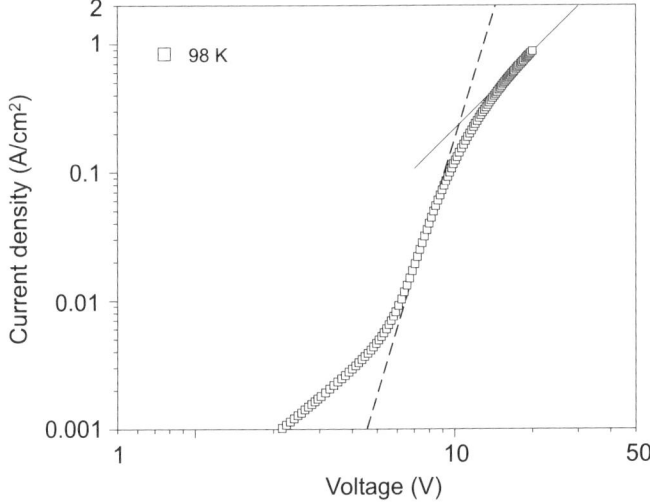

FIG. 3.22. $J-V$ characteristics of the conducting organic polymer diode ITO/PEDOT:PSS/MEH-PPV/Au at 98 K. The squares are the experimental data while the dotted line shows the trap-filling region and the solid line represents trap-filled V^2 region [53].

3.9.3. TEMPERATURE EFFECTS IN MEH-PPV. RECENT WORK

We prepared another diode with the same structure ITO/PEDOT:PSS/MEH-PPV/Au, but different thickness of MEH-PPV. The thickness of MEH-PPV in this sample was 65 nm. The experimental $J-V$ characteristics of this diode at 300, 243 and 210 K are shown by symbols in Fig. 3.23. As expected the conventional power law equa-

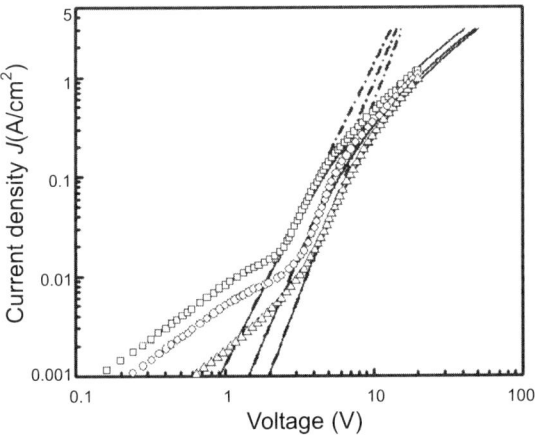

FIG. 3.23. Calculated and experimental $J-V$ characteristics of the ITO/PEDOT:PSS/MEH-PPV/Au diode at different temperatures with the active layer thickness of 65 nm. The symbols represent the experimental data, the dash-dot lines represent the conventional power law equation (3.42), while the solid curves represent the new theory equation (3.46) [44].

tion (3.42) (dash-dot lines in Fig. 3.23) holds well at the intermediate voltage range and as the voltage increases further the experimental results deviate from Eq. (3.42). The new theory including the effect of constant of integration C (Eq. (3.46)), at different temperatures are shown by the solid curves in Fig. 3.23. The experimental data agrees with the new theory quite well. The values of the parameters used for this sample are: $N_0 = 10^{19}$ cm^{-3}, $T_c = 600$ K, $\epsilon = 3$, $\epsilon_0 = 8.85 \times 10^{-14}$ F/cm, $N_v = 2 \times 10^{19}$ cm^{-3}, $H_b = 6.5 \times 10^{18}$ cm^{-3} and mobility $\mu = 3.6 \times 10^{-5}$, 6.5×10^{-5} and 9.5×10^{-5} cm^2/V s at 300, 243 and 210 K respectively. The mobility used here is the drift mobility, which is different from the trap-controlled mobility used in the mobility model discussed later.

Since the thickness of the sample used in Fig. 3.17 and the sample used in Fig. 3.23 are different, our theory holds for different temperatures as well as for different thickness of the samples.

3.9.4. TEMPERATURE EFFECTS IN ALQ3

Balanced charge carrier injection from electrodes into the active organic material at low voltages is essential for efficient operation of OLEDs. Ideally the anode and cathode should be able to form ohmic contact so that they can inject large density of charge carriers. In small molecular materials the injection of negative charge carriers is more efficiency-limiting factor rather than the hole injection. Thus a stable and efficient electron-injecting cathode is of great importance. Understanding of the charge transport in OLEDs is very complicated by the presence of both electrons and holes in working devices. A complete analytical model does not exist for such cases.

Researchers have used various models to explain the charge carrier transport mechanism in organic semiconductors. Two models have been used frequently, (i) the trapping model, which assumes a certain distribution of traps in the energy space and (ii) the field dependent mobility model, which assumes an exponential dependence of mobility on square root of electric field.

The transport properties of tris(8-hydroxyquinoline) aluminum (Alq$_3$) are of great interest [54]. Alq$_3$ is one of the most important emitters and among the electron transport materials used in OLED fabrication. Early measurements on Alq$_3$ [55] show that the transport at room temperature is by the trapping model, i.e. by the drift flow of electrons in the Lowest Unoccupied Molecular Orbital (LUMO). An activation energy of 0.1–0.2 eV was derived. The band model does not work at low temperatures because the available thermal energy is not sufficient to ionize the traps [42]. We measured the current in electron only devices (Al/LiF/Alq$_3$/LiF/Al) at different temperatures. After preparation the sample was directly transferred to the low temperature measurement assembly. It resulted exposure of the sample to the ambient atmosphere. This procedure is likely to introduce traps in Alq$_3$ [54]. We found that the charge carrier transport mechanism required to interpret the high temperature data is not sufficient to interpret the low temperature data. This result is in agreement with the results of Ref. [42].

3.9.4.1. Comparison of Experimental and Calculated J–V Characteristics of Alq$_3$

The observed J–V curves of Alq$_3$ electron only sample from 295 to 83 K are shown in Fig. 3.24. The band model with exponential distribution of traps fits well the data

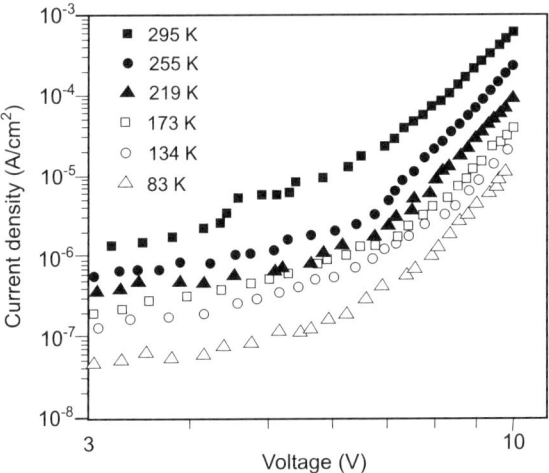

FIG. 3.24. $J-V$ characteristics of the Al/LiF/Alq$_3$/LiF/Al electron only device as a function of temperature for Alq$_3$ with a thickness of 160 nm [54].

at 295 K. The parameters used in the fitting are: $T_c = 2300$ K, $H_b = 2 \times 10^{17}$ cm^{-3}, $N_v = 3.4 \times 10^{16}$ cm^{-3}, $\epsilon = 3$, and $\mu = 1 \times 10^{-5}$ cm^2 V^{-1} s^{-1}. The theory did not match with the experimental data at lower temperatures if the same values of the parameters are used. However the data could be fitted with theory at 255 and 219 K by decreasing the mobility to 6.9×10^{-6} and 5.8×10^{-6} cm^2/V s respectively. The fits of experimental data with the theoretical curves at different temperatures with characteristic temperature $T_c = 2300$ are shown in Fig. 3.25. The value of trap density $H_b = 2 \times 10^{17}$ cm^{-3} was obtained by fitting the theory with the experimental curves. Using the same data the trap

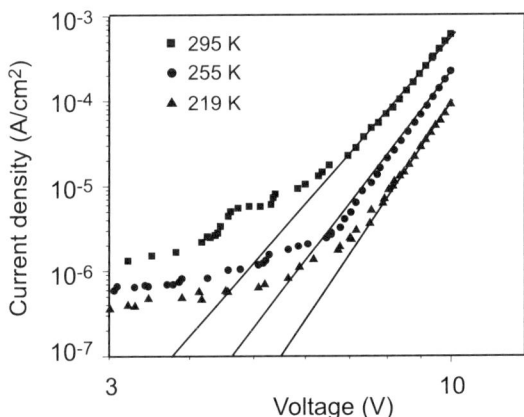

FIG. 3.25. Experimental and theoretical $J-V$ characteristics of the Alq$_3$ electron only device at higher temperatures. The solid lines are fits of the theoretical prediction to trap model where the current obeys the power law $J = V^{l+1}$, while the symbols are the experimental data. From these measurements characteristic temperature (T_c) of 2300 K can be derived [54].

density was also determined independently by the method used by Kumar et al. [49]. The value is comparable to that obtained here. We would like to point out that the mobility derived by fitting the trapping model to the experimental data is extrinsic apparent electron mobility due to traps in Alq$_3$, not the real intrinsic electron mobility. Measurements on intrinsic electron mobility of Alq$_3$ using different techniques are available in the literature. At lower temperatures, the trapping model did not fit the observed experimental data. It was possible to fit this model at lower temperatures if the values of H_b and T_c were changed. However it is not reasonable to assume that H_b and T_c are different at different temperatures. This result is similar to that already reported by Kapoor et al. [42] for poly(2-methoxy-5-(2-ethylhexyloxy)-1,4-phenylene vinylene) (MEH-PPV) as they used the mobility model at low temperatures. Our data fits the mobility model from 173 to 83 K with the same values of parameters. These results can be understood from general physical consideration. At room temperature sufficient number of electrons are ionized from the traps into the LUMO and the transport through the band model becomes appropriate. At lower temperatures, however, the number of electrons ionized in LUMO becomes small. Now the mobility model, which is predominantly controlled by hopping of charge carriers, becomes more important. We should like to point out that the transport by the band model and the mobility model occurs at all temperatures. It is their relative magnitudes that change in going from high temperatures to low temperatures. In the mobility model the J–V characteristics were obtained by solving numerically the couple Poisson and continuity equations. The theoretical J–V characteristics according to the mobility model with the experimental data at 173, 134, and 83 K are shown in Fig. 3.26. At high-applied bias the experimental data is in agreement with the theory. The values of the parameters used in the fitting of the mobility model are: $\mu_0 = 1 \times 10^{-15}$ cm^2/V s, $T_0 = 600$ K and the values of β (see Eq. (3.56) in the mobility model) used at different temperatures are: 5.5×10^{-4}, 4.2×10^{-4}, and 2.5×10^{-4} eV cm$^{1/2}$ V$^{-1/2}$ at 173, 134, and 83 K respectively. It is known that β increases as temperature decreases [54]. At first sight it may appear surprising that our values of β decrease as the temperature decreases.

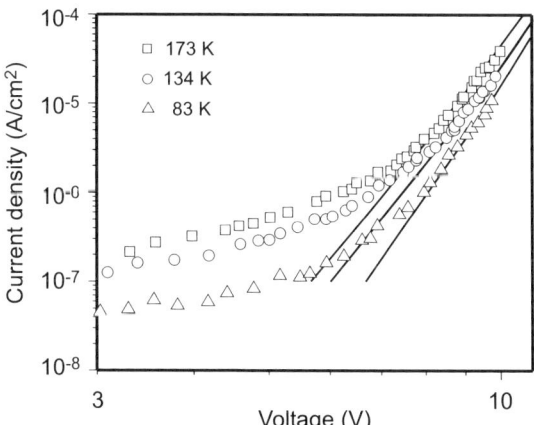

FIG. 3.26. J–V characteristics of the Alq$_3$ electron only device at lower temperatures. The solid lines are the theoretical plots while the symbols are the experimental data. From these results the current was found to obey the mobility model [54].

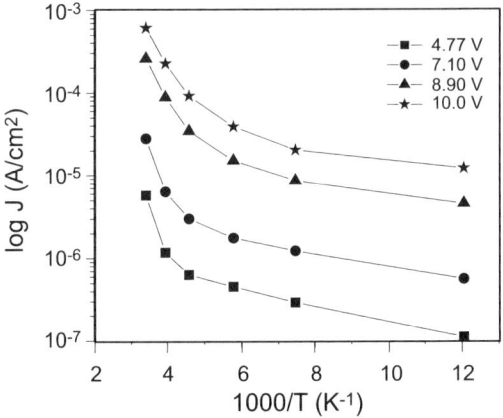

FIG. 3.27. Experimental values of J plotted as a function of $1000/T$ for $V = 4.77, 7.1, 8.9$ and 10 V for a 160 nm thick Alq$_3$ sample [54].

However a value of β decreasing at lower temperatures has also been reported. The value of β can decrease as temperature decreases due to screening of trapped charges. Fig. 3.27 shows the log J versus $1000/T$ plots at 4.77, 7.1, 8.9, and 10 V. The plots show two linear regions for each voltage, one at high temperature and the other at low temperature. The slope of the high temperature regime is larger compared to that in the low temperature regime. This large difference of the current increment in two regions suggests that the conduction mechanisms are different in the two regions. The origin of different mechanisms at high and low temperatures has been discussed in detail [7,23]. The activation energies at low temperatures were measured from the straight lines in low temperature region shown in Fig. 3.27. The values of the activation energies obtained are plotted in Fig. 3.28. The zero field activation energy $\Delta_0 = 0.05$ eV was obtained by the extrapolation of activation energy versus field$^{1/2}$ plot to zero fields.

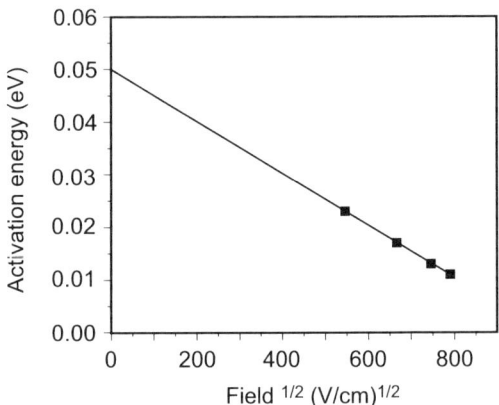

FIG. 3.28. Low temperature activation energy versus square root of electric field extrapolated to zero field. The zero field activation energy was found to be $\Delta_0 = 0.05$ eV [54].

We conclude that from the experimentally observed power law dependence of the current J on V^{l+1} with $l > 2$ at high temperatures the conduction is trap limited with an exponential distribution of traps having the characteristics energy $E_t = 0.19\,\text{eV}$. At high temperatures the conduction in the organic molecular solids is through the extended states. But at low temperatures there is not sufficient energy to ionize the traps and transport is governed by hopping of charge carriers through the localized states. At low temperatures the field dependent mobility model is in good agreement.

3.10. Mobility Model of Charge Transport

The mobility model is based on the empirical fitting of the time-of-flight data to obtain the observed mobility-field dependence [42, and references given therein]. In the mobility model the mobility μ is assumed to vary with electric field F according to the formula [46,40,56]

$$\mu = \mu_0 \exp(\beta\sqrt{F}/kT) \tag{3.56}$$

where

$$\mu_0 = \mu_{00} \exp(-E_t/kT) \tag{3.57}$$

and μ_{00} is a constant for a given material.[2] Eq. (3.56) has been used successfully for interpreting experimental results in a large number of materials and over a wide range of electric fields [56,40]. However a general derivation of this equation does not exist [57]. Extensive experimental work on the transport measurements in the hole only and electron only PPV LEDs has been done by Blom and collaborators (see their review, Ref. [57]). The result obtained with the electron only devices seemed to suggest that trapping of electrons determines the transport. To interpret results with the hole only samples it was found necessary to assume that the samples were free of traps but the mobility is field dependent. It is not clearly understood as to why the effect of disorder (which results in the field dependent mobility) is not present in the electron only devices. Generally traps are always present in all amorphous and disordered materials [38]. It was also not clear as to why the traps do not play any role in the transport of holes in the hole only devices. As pointed out recently in Ref. [41] the mobility model cannot explain the $J(V)$ relation over the whole range of applied voltages.

3.11. Unified Model

3.11.1. SHALLOW GAUSSIAN AND SINGLE LEVEL TRAPS

We treat here the case of shallow Gaussian traps. The equations reduce to those applicable to single level traps by making the standard deviation $\sigma_t = 0$. If the Poole–Frenkel effect (PFE) is included the J–V relation for a device containing shallow traps is given

---

[2] Different forms of this equation are given in Refs. [57] and [46].

by the following integral equation [38],

$$J = \frac{9}{8}\epsilon\epsilon_0 \mu_p \theta \frac{V^2}{d^3} \left\{ \frac{8}{9} \left[\int_{F(x=0)}^{F(x=d)} F X\, dF \right]^3 \left[\int_{F(x=0)}^{F(x=d)} F^2 X\, dF \right]^{-2} \right\}, \quad (3.58)$$

$$X = \exp[\beta \sqrt{F}/kT], \quad (3.59)$$

$$\theta = \frac{N_v}{g H_{dp}} \exp\left(-\frac{E_{tp} - E_v}{kT} + \frac{1}{2}\left(\frac{\sigma_t}{kT}\right)^2\right). \quad (3.60)$$

Here H_{dp} is the hole trap density, g is the degeneracy factor (taken as unity in the calculations), E_{tp} is the ionization energy of the hole traps, E_v is the valence band edge (i.e. HOMO), N_v is the effective density of states for holes, σ_t is the standard deviation of the Gaussian distribution of traps, and the factor $\exp(\beta \sqrt{F}/kT)$ arises due to the Poole–Frenkel effect. Analytical solution of (3.58) and (3.60) can not be obtained. Numerically computed results are shown in Fig. 3.29.

To derive Eq. (3.56) for the field dependent mobility we assume that the mobility $\mu(F)$ is an unknown function of F. We integrate the transport and Poisson equations and obtain,

$$J = \frac{9}{8}\epsilon\epsilon_0 \frac{V^2}{d^3} \left\{ \frac{8}{9} \left[\int_{F(x=0)}^{F(x=d)} \mu(F) F\, dF \right]^3 \left[\int_{F(x=0)}^{F(x=d)} \mu(F) F^2\, dF \right]^{-2} \right\}, \quad (3.61)$$

Eqs. (3.58) and (3.61) become identical if

$$\mu(F) = \mu_p \theta \exp(\beta \sqrt{F}/kT). \quad (3.62)$$

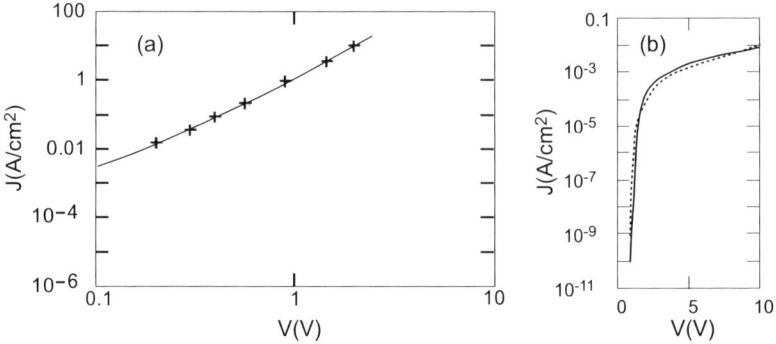

FIG. 3.29. (a) J–V values in ITO/PPV/Au are shown. The solid curve shows the calculated values, and the symbols show the 296 K experimental values from Ref. [57]. The value of $\mu_p \theta$ used is 5×10^{-7} cm^2/V s and $\beta/kT = 5.4 \times 10^{-3}$ (cm/V)$^{1/2}$. The values of the other parameters used are the same as in Ref. [57]. (b) The J–V curves of an Al/OC$_1$C$_{10}$/ITO Schottky diode are shown. The solid line shows the calculated values assuming exponentially distributed traps. The dotted curve shows the experimental data. The values of the parameters used in the calculations are: $H_{dp} = 1 \times 10^{18}$ cm^{-3}, $N_v = 3 \times 10^{19}$ cm^{-3}, $d = 130$ nm, $T_c/T = 10$, $T_c = 3000$ K, $\mu = 1 \times 10^{-6}$ cm^2/V s, $\beta_{PF}/kT = 0.011$ (cm/V)$^{1/2}$ and $\epsilon = 3\epsilon_0$ (ϵ_0 is the dielectric constant of free space) [56].

Eq. (3.56) for the field dependent mobility and Eq. (3.62) are identical if

$$\mu_{00}\exp(-E_t/kT) = \mu_p\theta.$$

If we consider a sample with shallow Gaussian traps and include PFE, the sample behaves as if there are no traps and the mobility is field dependent given by Eq. (3.56) far as the dependence of J on V is concerned. The zero field mobility and its temperature dependence are different in the two equations. If the traps are at a single energy level, $\sigma_t = 0$ and the temperature variation of the mobility also becomes the same in the two cases. Eq. (3.58) represents both the models, it reduces to the existing shallow trap model (without PFE) when $\beta = 0$ and to the existing field dependent mobility model when $\theta = \exp(-E_t/kT)$.

Fig. 3.29(a) shows that the results of the unified model (based on shallow traps and with PFE) agree well with the 296 K J–V curve measured by Blom et al. [57]. Blom et al. [57] found excellent agreement with the theory that does not assume the presence of traps but uses the field dependent mobility. The above discussion shows that the model with shallow traps plus PFE and the field dependent mobility model are equivalent.

3.11.2. UNIFIED MODEL WITH EXPONENTIALLY DISTRIBUTED TRAPS

Fig. 3.29(b) shows the published J–V curves of an Al/OC$_1$C$_{10}$/ITO diode by Jain et al. [40]. [OC$_1$C$_{10}$ is poly(2-methoxy-5-(3,7-dimethyloctyloxy)-p-phenylene vinylene).] In this paper [40] Jain et al. attributed the extremely fast rise of the hole current at low voltages to Shockley like current due to the forward biased Al Schottky contact. In their later paper [56] Jain et al. showed that by including the PFE they can fit the theory of bulk limited trap controlled space charge currents with the same experimental results (see Fig. 3.29(b)). Jain et al. [40] suggested that PFE induces the high injection effect earlier, which presumably makes the Schottky barrier at the Al contact small.

Jain et al. [40] fabricated new organic diodes using five-ring PPV-type oligomer 2-methoxy-5-(2'-ethylhexyloxy)-1,4-bis((4',4"-bisstyryl)styrylbenzene) (MEH-OPV5, see its structure in Fig. 3.30). The MEH-OPV5 films were sublimed in UHV (10^{-9} Torr) at a source temperature of 210 °C on a cooled ITO coated glass substrate. The metals were evaporated through a shadow mask. The J–V characteristic of the ITO/MEH-OPV5/Al diode is shown by the solid curve in Fig. 3.30. The effect of an interfacial layer at the ITO electrode was also investigated. A device with an interfacial layer of PEDOT inserted between the ITO and the organic layer was fabricated and its J–V curve was measured (see the dotted curve in Fig. 3.30). Both solid curve (no interfacial layer) and dotted curve (interfacial layer between ITO and the organic layer) show the rapid rise of the current at small voltages. This behavior is similar to that of the earlier diode shown in Fig. 3.29 and can be explained similarly, i.e. by either of the two mechanisms discussed above. The behavior of the J–V characteristics observed in the voltage range between 0 and 1 V is ohmic. It is attributed partly to shunting paths between both electrodes and partly to background doping. The shunting paths can originate from pinholes in the MEH-OPV5 layer. Covering the ITO electrode with a smoothening PEDOT layer suppresses the ohmic currents due to shunting paths. The remaining low voltage current is probably due to the background doping. Measurements of J–V characteristics at different temperatures

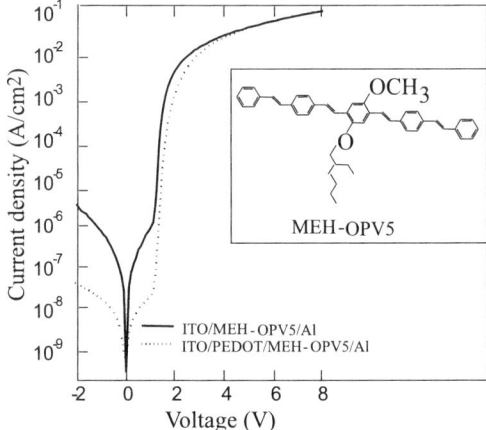

FIG. 3.30. J–V curves of two MEH-OPV5 based diodes. The structure of MEH-OPV5 is shown in the inset [56].

should help determine the correct mechanism of current transport in Al contacted organic diodes.

High field has a large effect on the J–V characteristics of conducting organic materials and PFE must be included in the calculations. When PFE is included in the calculations, most experimental results interpreted in the literature by the mobility model can be explained by the trapping model. This has great advantage because physics of the trapping model is well understood. The fast rise of the current at small voltages can be explained by the Schottky barrier at the Al contact or by assuming the Poole–Frenkel detrapping and exponentially distributed traps. In Fig. 3.29 we have adjusted the values of β and T_c to fit the theory with the experiment. Unless reliable values of T_c become available from other independent experiments, it is difficult to determine quantitatively the relative importance of the PFE in this case.

3.12. High Field or Poole–Frenkel Effect

3.12.1. J–V Characteristics

If the electric field is high, ionization energy of the traps is reduced by the Poole–Frenkel Effect (PFE) [38, and references given therein] or the Arnett effect [58]. The relation (3.42) between current and voltage is also modified. The Poole–Frenkel effect is illustrated in Fig. 3.31. The reduction in energy is given by $\beta_{PF} F^m$ where β_{PF} is the so-called Poole–Frenkel constant [38] and m varies between 0.5 and 1 depending on the nature of traps. We use the Poole–Frenkel value $m = 0.5$. The reduction in the trap depth is given by the following equation [38],

$$E_{PF} = \beta_{PF}\sqrt{F}. \tag{3.63}$$

If screening is taken into account, the value of this coefficient is modified [39].

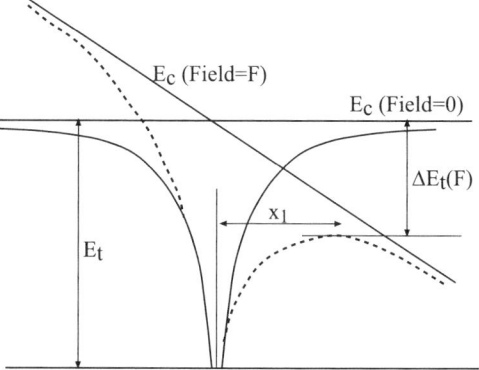

FIG. 3.31. Schematic illustration of the lowering of trap ionization energy by the Poole–Frenkel effect.

We now apply the PFE to the transport of charge carriers in a polymer containing traps at a single energy level. In the presence of high electric field, the energy appearing in the denominator of the Fermi function in Eq. (3.24) is reduced by $\beta_{PF}\sqrt{F}$. The energy in the δ function remains unchanged. The density of the trapped holes is reduced and is given by,

$$p_{t} = \frac{H_{a}}{1 + (\theta_{a}H_{a}/p(x)) \exp[\beta_{PF}\sqrt{F}/kT]}. \tag{3.64}$$

This can be added on the right-hand side of Eq. (3.4). The modified Poisson equation coupled with the transport equation (3.5) can be solved to yield,

$$\int_{0}^{d} dx = \int_{F(x=0)}^{F(x=d)} \frac{\epsilon\epsilon_{0}}{q} (F_{1})^{-1} dF, \tag{3.65}$$

where the function F_1 is defined as,

$$F_{1} = \left[\frac{J}{q\mu_{p}F} + \frac{H_{a}}{1 + \theta_{a}H_{a}q\mu_{p}F \exp(\beta_{PF}\sqrt{F}/kT)/J}\right]. \tag{3.66}$$

For a given J and d, Eq. (3.65) is solved numerically to obtain the field $F(d)$ at $x = d$. The field is assumed to be zero at $x = 0$. The field $F(d)$ is now divided in n equal parts $F_1, F_2, \ldots, F_n$. Corresponding values $x_1, x_2, \ldots, x_n$ are determined using Eq. (3.65) again. Since $F(x)$ is now known as a function of x, the total voltage drop across the sample (which is equal to the applied voltage) is obtained by integrating Eq. (3.6). This procedure is repeated for several values of J and complete J–V characteristics are obtained.

In the case of exponentially distributed traps, the effect of high fields on the trap depths (Poole–Frenkel effect) can be taken into account by the same procedure. The trapped hole density is modified and Eq. (3.40) changes to [38],

$$p_{t} = \int_{E_{FP}}^{\infty} \frac{H_{b}}{kT_{c}} \exp\left(\frac{-E}{kT_{c}}\right) \frac{1}{1 + \exp[(E_{F} - [E - \beta_{PF}\sqrt{F}])/kT]} dE. \tag{3.67}$$

This equation gives the final expression for trapped holes in the presence of high field effects,

$$p_\text{t} = H_\text{b} \exp\left(\frac{-\beta_\text{PF}\sqrt{F}}{kT_\text{c}}\right)\left[\frac{p}{N_\text{v}}\right]^{1/l}. \quad (3.68)$$

Eq. (3.68) shows that high field reduces the number of trapped carriers significantly. Now an analytical solution corresponding to Eq. (3.42) cannot be obtained. The solution can be written in the following form,

$$\int_0^d dx = \frac{\epsilon\epsilon_0}{q} \int_{F(x=0)}^{F(x=d)} \frac{dF}{F_2}, \quad (3.69)$$

where

$$F_2 = \frac{J}{q\mu_p F} + H_\text{b} \exp\left(\frac{-\beta_\text{PF}\sqrt{F}}{kT_\text{c}}\right)\left[\frac{J}{q\mu_p F N_\text{v}}\right]^{1/l}. \quad (3.70)$$

This equation is solved numerically by the method described earlier for the single level traps.

3.12.2. Calculations and Comparison with Experiments

Kumar et al. [39] made numerical calculations for traps at a single energy level and for traps distributed exponentially in the energy space. The effect of high field is qualitatively similar in the two cases. We show the effect of high field on the electric field in Fig. 3.32. The electric field and Poole–Frenkel effect suppress the actual electric field considerably. Near the exit end the field is suppressed by more than one order of

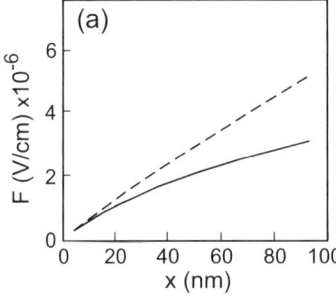
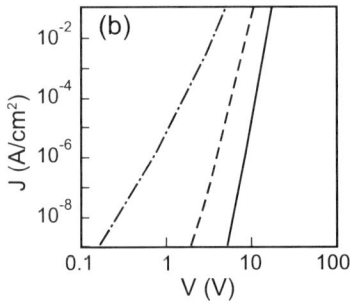

FIG. 3.32. (a) Calculated electric field versus distance in a conducting polymer containing exponentially distributed traps for $J = 5 \times 10^{-5}$ A cm^{-2}. The values of the parameters used in the calculations are: $T_\text{c} = 3000$ K, $H_\text{b} = 3 \times 10^{18}$ cm^{-3}, $d = 100$ nm, $N_\text{v} = 3 \times 10^{19}$ cm^{-3}, and $\mu = 5 \times 10^{-5}$ cm^2/V s. The dashed curve is without PFE and the solid curve is with PFE. (b) Calculated $J(V)$ curves in polymers with exponential trap distributions. The values of T_c are 3000 K (solid curve), 2000 K (dashed curve) and 1000 K (dash-dot curve). The values of the other parameters are the same as in (a).

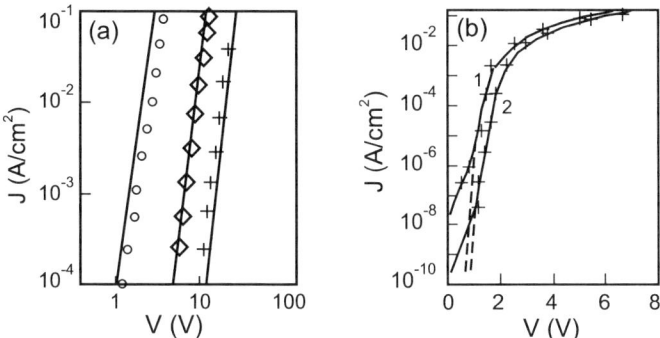

FIG. 3.33. (a) Comparison of the FDTO model (solid lines) with the experimental (symbols) of Crone et al. [59] for a Ca/MEH-PPV/Ca electron only device. (b) Comparison of the FDTO model (dashed line) and FDTO *with background impurity* model (solid curve) with the experimental data for Al/MEH-OPV5/ITO (curve 1) and for Al/MEH-OPV5/PEDOT/ITO (curve 2). The thickness of both samples is 110 nm.

magnitude. The suppression is even larger for the single level traps (see Fig. 1(a) of Ref. [39]).

According to Eq. (3.5) for a given J the product of p and F remains constant. Suppression of F results in a large increase in the carrier density. Therefore the current is expected to increase due to PFE. Numerical calculations show that there is a large increase in the current when PFE is switched on (see Fig. 2 of Ref. [39]).

Several experimental $J-V$ curves taken from the recent literature have been compared with the PFE model [39]. We quote a few typical examples here. Extensive measurements of $J(V)$ curves have been made by Crone et al. [59]. They used both *hole only* Au/MEH-PPV/Al and *electron only* Ca/MEH-PPV/Ca devices. Crone et al. [59] used the mobility model to interpret their results. Comparison of their results for the electron devices with the Field Dependent Trap Occupancy (FDTO) model is shown in Fig. 3.33(a). The thicknesses of the samples are 25 nm (circles), 60 nm (diamonds) and 100 nm (plus symbols). The parameters used in calculations are: $T_c = 1420$ K, $H_b = 7 \times 10^{18}$ cm^{-3}, $N_t = 5 \times 10^{19}$ cm^{-3}, $N_v = 1 \times 10^{20}$ cm^{-3}, $\mu = 6.5 \times 10^{-6}$ cm^2 V^{-1} s^{-1} and $\epsilon = 3$. The agreement of the experimental results is satisfactory. The model has another advantage. The calculations have been made with the same parameters for all the three samples. With the mobility model, Crone et al. [59] had to use different values of the parameters for different samples.

Comparison of the FDTO model with the experiments performed at IMEC is shown in Fig. 3.33(b). Two sets of diodes were fabricated: (1) Al/MEH-OPV5/ITO (curve 1) and (2) Al/MEH-OPV5/PEDOT/ITO (curve 2) [40,56]. The values for fitting parameters used in calculations are: $H_b = 8 \times 10^{16}$ cm^{-3}, $\mu = 2 \times 10^{-5}$ cm^2 V^{-1} s^{-1}, $N_v = 1 \times 10^{20}$ cm^{-3}, $T_c = 4000$ K, $\epsilon = 2$ and back ground doping $p_0 = 9 \times 10^{11}$ cm^{-3} (curve 1) and $H_b = 1.1 \times 10^{17}$ cm^{-3}, $\mu = 1.8 \times 10^{-5}$ cm^2/V s, $N_v = 1 \times 10^{20}$ cm^{-3}, $T_c = 4000$ K, $\epsilon = 2$ and back ground doping $p_0 = 7 \times 10^{10}$ cm^{-3} (curve 2). Fig. 3.33(b) shows excellent agreement of the FDTO model with the experiment provided effect of back ground impurity is taken into account.

3.13. Mobility of Charge Carriers

3.13.1. BULK MATERIALS

In a drift dominated case the recombination probability depends on the mean drift length $W_{h/e} = \mu_{h/e} \tau_{h/e} F$. If the mean drift length of one or both of the charge carriers is smaller than the device length d the recombination is significant. The saturated photocurrent density is given by [60],

$$J_{ph}^{sat} = qgd. \tag{3.71}$$

The hole mobility in MDMO:PPV is 5×10^{-11} m^2/V s and the electron mobility in PCBM (a derivative of C_{60}, see Fig. 3.1) is 2×10^{-7} m^2/V s. Due to large difference in the mobilities of the two charge carriers the holes accumulate in the device. This increases the electric field near the anode which aids the exit of the holes. The electrostatic limit of the accumulation is reached when the photocurrent density reaches the space charge limited current density. Melzer et al. [60] found that if the above value mobility is used, the observed current exceeds the predicted limit. The only way to understand this result is to accept that the mobility in the blend is more than in the MDMO:PPV alone. Melzer et al. [60] measured the hole mobility in the 1:4 weight ratio blend of MDMO:PPV and PCBM by three methods, the current–voltage measurements, transient electroluminescence (EL) measurements and admittance measurements. The results of these measurements are shown in Fig. 3.34(a). The results obtained using three quite different techniques agree closely. More importantly the hole mobilities in the alloys are some two orders of magnitude larger than those in the neat MDMO-PPV. They are now much closer to the electron mobility in the PCBM. The possible mechanisms by which this increase can be obtained are shown in Figs. 3.34(b) and 3.34(c).

Hewig and Bloss [61] and Agostinelli et al. [62] have observed the same phenomenon in the chalcogenide solar cells. At higher voltages the junction current measured in the dark becomes smaller than the solar cell output current under illumination. Agostinelli et

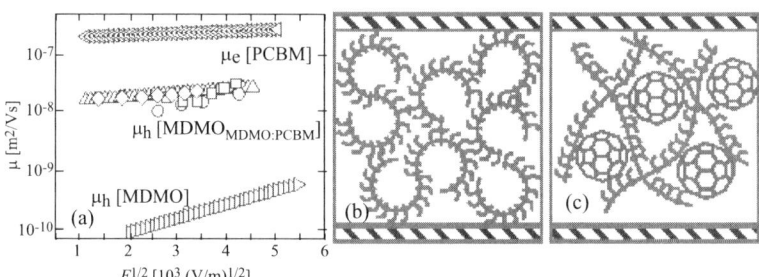

FIG. 3.34. (a) The charge carrier mobilities in the PCBM [51], MDMO [63], and MDMO:PCBM blends [60]. Only the blend mobilities were determined by measurement of space charge limited currents (△), by admittance spectroscopy (◇), and by transient measurements of a 200 nm (□) and a 120 nm (○) layer of the blend and 60 nm MEH-OPV5. (b) The molecular conformation of neat MDMO:PPV according to Kermerink [64]. (c) Upon mixing the MDMO:PPV with PCBM the interactions between the two moieties might change leading to the new conformation.

al. [62] have discussed mechanisms that can modify the junction current flow on illumination.

3.13.2. MOBILITY IN BLENDS

The hole mobility μ_h in OC_1C_{10}-PPV is 5×10^{-11} m^2/V s. The electron mobility μ_e in PCBM is 2×10^{-7} m^2/V s. However, the transport of separate charge carriers in an interpenetrating network may be different than the transport in the individual compounds. In Ref. [65] the transport and injection of charge carriers in OC_1C_{10}-PPV:PCBM bulk-heterojunction diodes are investigated.

The J–V characteristics of an ITO/PEDOT:PSS/OC_1C_{10}PPV:PCBM (1:4 w/w)/LiF/Al diode and electron current in a PCBM device have been measured. The thickness of each device was $L = 170$ nm. The work function difference between the ITO/PEDOT:PSS and LiF/Al contacts gives rise to a built in voltage of 1.4 V. The applied bias V was corrected for the built-in voltage. At large current densities the applied voltage was corrected for the voltage drop across the series resistance at the ITO contact. Typically the series resistance was 11 Ω. The slope of the log J–log V plot shows that the current density depends quadratically on the voltage (J proportional to V^2), consistent with the space charge limited current SCLC. In forward bias the dark current in OC_1C_{10}-PPV:PCBM (1:4 w/w) bulk heterojunction diodes is equal to the electron current in pure PCBM. This shows that the electron mobility of PCBM in the bulk heterojunction is not appreciably different from that in pure PCBM. The SCLC behavior of PCBM current demonstrates that the use of LiF/Al cathode does not represent a significant barrier for electron injection.

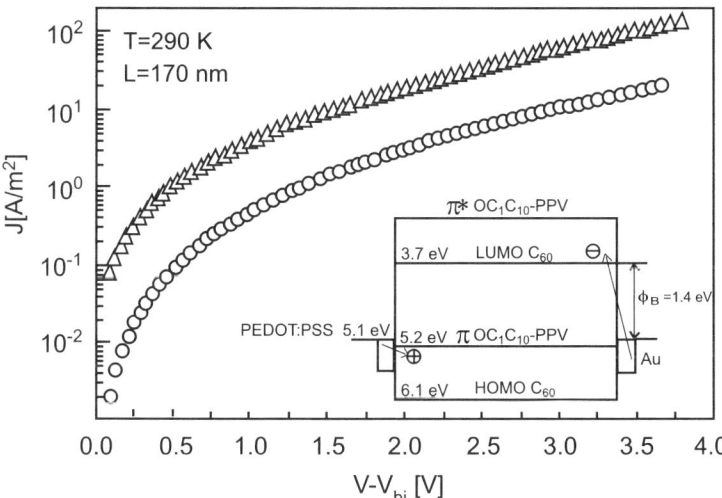

FIG. 3.35. Experimental J–V characteristics of an ITO/PEDOT:PSS/PCBM/Au injection limited electron current (triangles) and calculated space charge limited hole current in OC_1C_{10}-PPV (circles) for a thickness of $L = 170$ nm and temperature $T = 290$ K. The inserted figure represents the device band diagram under the flat band condition of a bulk heterojunction solar cell using Au as a top electrode [65].

The measured electron current in PCBM injected through Au electrode is shown in Fig. 3.35. Calculated SCL hole current in OC_1C_{10}-PPV is shown by circles in Fig. 3.35 for a thickness of $L = 170$ nm. Even with the high Schottky barrier of 1.4 eV (the barrier determined from the measured current is 0.76 eV) with Al cathode for PCBM is not sufficient to suppress the electron current in PCBM below the hole current in OC_1C_{10}-PPV. With the large injection barrier the electron current is injection limited. This work shows that it is not possible to make hole only devices using OC_1C_{10}-PPV:PCBM bulk heterostructures.

3.14. Important Formulas

We give below formulas which are frequently needed in modeling the transport in polymers.

Tunneling currents according to the Fowler–Nordheim theory are given by, [50]

$$J = (C/\phi_B)(V/d)^2 \exp\left[-B\phi_B^{3/2}/(V/d)\right].$$

Pure ohmic conduction is given by

$$J = qP(0)\mu V/d.$$

This is Eq. (3.49). Pure space charge limited current with no traps is given by

$$J = (9/8)\epsilon_r\epsilon_0\mu V^2/d^3$$

where $\epsilon_r\epsilon_0$ is the dielectric permittivity. This is Eq. (3.12).

Space charge limited current in the presence of exponential distribution of traps is given by,

$$J = \frac{\mu_p N_v}{q^{l-1}} \left(\frac{2l+1}{l+1}\right)^{l+1} \left(\frac{l}{l+1}\frac{\epsilon\epsilon_0}{H_b}\right)^l \frac{V^{l+1}}{d^{2l+1}}.$$

This is Eq. (3.42). If the barrier is not zero, the current is given by,

$$J = \frac{\mu_p N_v}{q^{l-1}} \left(\frac{2l+1}{l+1}\right)^{l+1} \left(\frac{l}{l+1}\frac{\epsilon\epsilon_0}{H_b}\right)^l \frac{V^{l+1}}{\left[(d+C)^{\frac{2l+1}{l+1}} - C^{\frac{2l+1}{l+1}}\right]^{l+1}}.$$

This is Eq. (3.46).

The Schottky barrier in polymers is very different from the barriers in Metal-Si. Earlier computer simulations showed that in n+/n− junction no depletion is formed, instead accumulation is formed. Also if d is small, in n+/p− junction, there are not enough carriers to bring the Fermi levels in one line. In numerous cases analytical solutions give erroneous results and create understanding. Extensive computer simulations are required.

3.15. Summary of This Chapter

We started with a discussion of optical properties. The discussion is brief and covers points relevant of all polymers. The optical properties of individual polymers are discussed in the following chapters. We have given a detailed account of the transport properties. Greatest advances have been made in this area. the current–voltage ($J-V$) characteristics of poly(2-methoxy-5-(2-ethyhexyloxy)1,4-phenylene vinylene) (MEH-PPV) have been studied in the hole only devices ITO/PEDOT:PSS/MEH-PPV/Au, as a function of temperature from 300 to 98 K. Hole conduction in MEH-PPV has been well explained by the exponential trap controlled drift model. In a considerable range of applied voltages, current obeys the V^{l+1} law, as the voltage increases further the current deviates and becomes smaller. At low temperatures and sufficiently high-applied voltages, current follows the V^2 law. The polymer poly(p-phenylene vinylene) (PPV) and its derivatives are some of the most important materials for the fabrication of the polymer light emitting diodes (PLEDs) and polymer photovoltaic devices. Performance of all these devices depends primarily on the electrical conduction property of the polymers. The conduction mechanism in organics is a subject of great topical interest. For further development of these devices, understanding of the electrical conduction mechanism in such materials is of great importance.

CHAPTER 4
LIGHT EMITTING DIODES AND LASERS

4.1. Early Work

A respectable electroluminescence[1] from an organic molecular solid device was first demonstrated by Kodak in 1987 [66]. The first polymer LED was fabricated by Cambridge University in 1990 [67]. This stimulated a lot of excitement and intense activity in the field of organic electronic and optoelectronic devices. Considerable work is now being done on the Organic Solar Cells, Organic Thin Film Transistors (OTFTs), Organic Light Emitting Diodes (OLEDs) and lasers.

C.W. Tang and collaborators fabricated the first organic LED in 1987 using tris(8-hydroxyquinoline) aluminum (Alq$_3$) emitting green light. The chemical structure of Alq$_3$ and other organics used in the LED technology is shown in Fig. 4.1. The LED efficiency, in lumens per watt, can be more than that of the household bulb and their life can be more than 10,000 hours. In 1994 Ananth Dodabalapur and his co-workers at Bell laboratories constructed EL devices by sandwiching Alq$_3$ between two reflecting surfaces forming a microcavity. This structure conforms to the physics of Fabry–Perot cavity. In the conventional structures, the light is wasted since it leaks in all directions. By varying a thickness of the inert layer, undesirable light can be filtered out and emission of light can be obtained at any desired wavelength. Since a microcavity LED is more efficient and uses less current, it lasts longer. Different polymers emit different colors. Absorption edge in a large number of polymers has been measured. The typical values are: 1.6 eV for PANI, 2 eV for PT, 2.4 eV for PPV, 2.4 for PPy, and 3.0 eV for PPP. The mobility values for the polymers vary between 0.01 to 10^{-7} cm^2/V s. The chemical structure of the PPV, CN-PPV and junction LED is shown in Fig. 4.2. The group at the Cavendish laboratory headed by R.H. Friend [70, and references given herein] used PPV and its derivatives for fabricating the LEDs. They observed the green–yellow emission from the PPV LED in 1990. These LEDs gave 2.5 lumens per watt. If driven at high volts light output increases but faster breakdown of the devices occurs because of the heat generated. The competition with existing devices is going to be tough. LEDs for indicator lamps are cheaper. Characteristics of a typical organic LED fabricated by Li et al. [68] are shown in Fig. 4.3.

At present the GaAs wafer size is limited to about 6 inch diameter. To make large displays, GaAs based LEDs must be individually mounted and wired. A reasonable letter size can take up to 35 LEDs. The size of organic films is practically unlimited. Moreover the starting material for organic displays is much cheaper. Intense effort has been made in developing the thin film organic light emitting diodes [68, and references given therein].

At Siemens the effort is directed to develop devices for use in backlighting for the keys on mobile telephones or for the liquid crystal display devices. Polymers displays

---
[1] This chapter is based on an review written by Pankaj Kumar and collaborators.

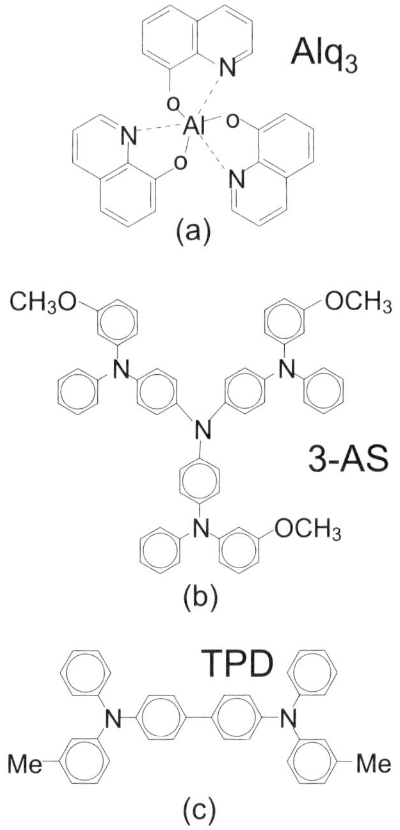

FIG. 4.1. Molecular structure of (a) 8-tris-hydroxyquinoline Al (Alq$_3$), (b) "three-armed" star (3-AS) 4,4',4"-tris[N-(3-methoxyphenyl)-N-phenylamine-triphenylamine], and (c) N,N'-diphenyl-N,N'-bis(3-methylphenyl)-1,1'-biphenyl-4,4'diamine (TPD) [68].

have an edge on LCD displays as it is hard to see these displays from an angle. Philips at Eindhoven has had a research program on polymers for several years. They have an agreement with Cambridge Display Technology (CDT) which gives Philips access to CDT's patented technology. CDT is a spin-off company of the research group at the Cambridge Cavendish Laboratory set up in 1992 by R. Friend and A. Holmes of Cambridge University. The group has also fabricated the PPV LEDs with microcavities. They have demonstrated optically pumped lasing in these structures. CDT believes that over a period of time cathode ray tubes could be replaced by the polymer devices. The display business is estimated to be more than $42 billion shortly. Polymer scientists are trying hard to carve a share for themselves from this business [69]. Large polymer screens can be easily produced by sandwiching three or four layers of polymers between two electrodes. Pixels can be created by patterning the electrodes. In May of 1996, CDT gave a public demonstration of a dot matrix display at the Society for Information Display meeting in San Diego. Other products which can benefit from polymer displays are Personal

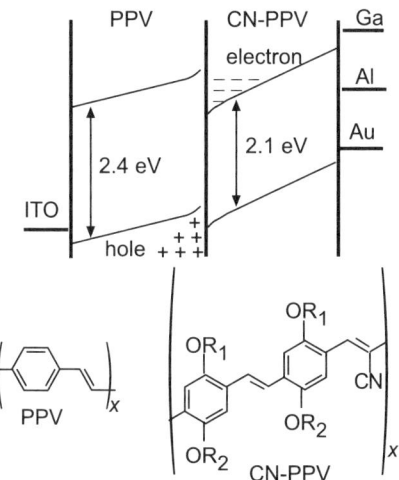

FIG. 4.2. Chemical structure of PPV, CN-PPV and of a PPV:CN-PPV junction LED [69].

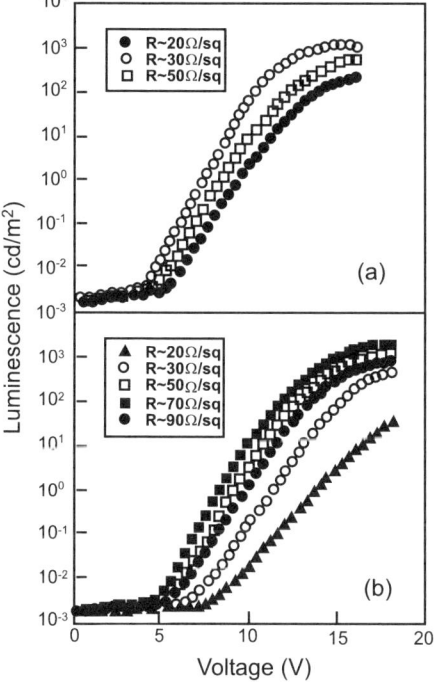

FIG. 4.3. Characteristics of a typical organic LED fabricated in 1997 [68].

Digital Assistants and organizers, CD players, electric razors, alarm clocks, radios and televisions.

Many organic lasers consist of a transparent gain medium doped with a suitable dye. The volume fraction of the dye is about 1% so that the spacing between dye molecules is sufficiently large to avoid concentration quenching. The transparent medium plays no part in the laser action except to support the dye molecules. The thickness of the film is from 150 to 200 nm so that it supports only the lowest-order optical mode. Because of the small quantity of the dye present in this thickness, the threshold power needed for stimulated emission is large. This difficulty can be avoided by using conjugated polymers or small molecules which have optical gain [71]. However with such structures a large part of the emitted light is absorbed in the host medium itself which raises the threshold power.

Berggren et al. [71] have used 2-(4-biphenylyl)-5-(4-t-butylphenyl)-1,3,4-oxadiazole (PBD) which is doped with about 1 wt% of fluorescent dyes with different absorption and emission spectra. The dyes used are C490, DCM2, and LDSB821. In these structures the emission wavelengths are very different from the absorption wavelengths. The host medium absorbs the pump light and the excitation is funneled to the dye medium by the so-called Förster energy transfer process [71]. The average space between the dye molecules is kept at about 3 nm so that the transfer of energy is efficient. The samples were pumped with $\lambda = 337$ nm pulsed unfocused light from a nitrogen laser. At low input power the PL from PBD and from the dyes was broad. When the input power was increased the peaks became narrow. However the threshold input power was rather large. Berggren et al. [71] also studied PBD doped with conjugated polymer, a soluble derivative of poly(phenylene vinylene) (10 wt%) and an oligomeric phenylene vinylene (2 wt%). In both cases the amplification of the emission was observed at a low input power of 200 W cm^{-2}.

4.2. Blue, Green and White Emission

4.2.1. Blue and Green LEDs

The OLEDs are generally fabricated on an indium-tin-oxide (ITO) coated glass plate. Recent work has shown that films grown by vacuum evaporation have considerably better quality. The cathode is a low work function metal or alloy such as Ca, Mg:Ag and Al:Li. Electrons and holes are injected from the cathode and the ITO anode respectively. An organic layer with good electron transporting and hole blocking properties is used next to the cathode. Similarly a hole transporting layer and electron blocking layer is used next to the anode. The electrons and holes injected by the cathode and the anode are driven to the middle of the structure where the electron and hole transporting layers meet. This produces a configuration similar to the quantum wells in the inorganic devices. The electron and holes are confined in a small region at the junction of the two layers. The probability of recombination and emission of light increases. Organic blue and green light emitting devices have been fabricated using three material systems, (1) small molecules, (2) conducting polymers and (3) organic matrices doped with dyes. Burrows et al. [72] have fabricated blue, green and red light emitting diodes for a full color display. Small

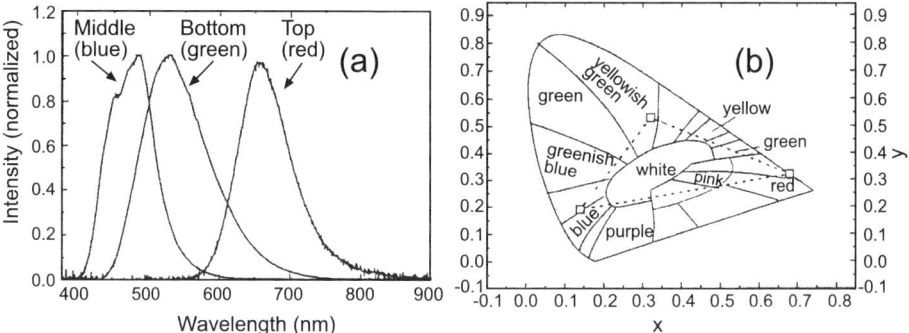

FIG. 4.4. (a) Spectra of the blue, green and red LEDs. (b) The spectra in (a) plotted on the standard CIE color chart illustrating the range of available color tuning [72].

molecular organic Alq$_3$ was used as a green emitter, Alq$_2$Oph as a blue emitter and 3% DCM2 dye doped Alq$_3$ as a red emitter. α-NPD was used for the hole transporting layers. The spectra of the three LEDs are shown in Fig. 4.4(a). The three diodes were vertically stacked over each other with transparent electrodes inserted in-between. Each LED was independently addressable. The available colors are illustrated in the CIE color chart in Fig. 4.4(b). The spectra shown in Fig. 4.4(a) are of the stacked devices. The spectra of the stacked devices are somewhat different from those of individual unstacked devices because of the cavity interaction and coupling effects. The 1.0 mm diameter devices operated at 0.1–0.5 mA and 15–25 V. The measured external quantum efficiencies are 0.4% for the green, 0.06% for the blue and 0.05% for the red LED. Red and blue light had to pass through a Mg:Ag electrode which had only 50% transmission. External efficiency of 1% is expected from the optimized design of the LEDs in the stacked structure.

Among the polymers PPV, MEH-PPV, BuEH-PPV, DOO-PPV and Si-PPV have been used to fabricate LEDs and optically pumped lasers. Most of these polymers emit the light in the yellowish–green to red regions of the spectrum [73].

A full color display can be fabricated in different ways such as patterning the pixels individually for three principal colors blue, green and red, using efficient dyes to convert colors [74], color tuning by chromaticity tuning layer [75] or applying different bias potentials to LEDs, using white OLEDs and filtering white light for a specific color. The filtration of white light is simple but there is much wastage of energy due to the absorption of unwanted light. Voltage tunable performances of OLEDs based on some materials, which produce different colors depending on the operating voltage, have been reported [76]. However, in such devices it is not easy to control the light intensity and the emission color at the same time. As the energy of a shorter wavelength can be converted into the energy of a longer one, using proper dyes blue light can be used to generate low energy emission of green or red. A blue OLED alone has the potential to generate all the colors, while the converse: conversion of green or red emission into blue, is not possible. Generally, the blue OLEDs have lower photometric efficiencies than the green or red OLEDs. Moreover human eyes have lower sensitivity in the blue region of visible spectrum. Although all the three principal colors have been demonstrated in OLEDs, only green and orange OLEDs currently have the abilities to meet the requirements for commercial applications. Efficient and stable organic blue OLEDs are still uncommon. The display technology using OLEDs is seriously handicapped from the lack of efficient and stable blue emitting materials. Development of blue OLEDs which can be produced commercially is a subject of current interest [77]. The red OLEDs should also be improved further. Design of stable and highly efficient blue emitters is a long-term research and development goal.

Blue emission has been studied in a number of organic materials like 1,3,4-oxadiazole derivatives [78], spiro-bifluorene derivatives [79], metal chelates such as boron, aluminum, lithium, gallium, silicon, beryllium complexes [80], amine derivatives [81], 1,8-acridinedione derivatives [82], anthracene derivatives [83] and polymers like PPV, fluorenes, thiophene and pyridine derivatives [52]. Polymer Light Emitting Diodes (PLEDs) are growing as the main display system because of many advantages concerning preparation and operation over other display systems. PLEDs exhibiting quantum efficiency up to 10% and 20 lm/W with lifetime of 20,000 h [70] have been reported. The most important merit of PLEDs is their easy fabrication. Polymer films can be easily deposited by spin coating their solution in appropriate solvents. In addition the ink-jet printing technique is a promising technology for patterning the pixels of primary colors. The technique is already available in top quality inject printers. An LED with a cyano-containing PPV based copolymer (PC10) as an emissive layer has EL peak in the blue region at 493 nm [84]. A dioctyl-substituted polyfluorine has the EL peak at 436 nm. Among the dyes Coumarin 503 dissolved in different polymeric matrices gives emission in the blue region. In PMMA, MMA and in copolymers MMA and styrene the peak wavelengths are in the range 472–494 nm [85]. The EL peaks in CBP doped with dyes are 485 nm for perylene, 460 nm for Coumarin 47 and 510 nm for Coumarin 30. Another polymer, which has been used by several groups for fabricating blue emitters, is LPPP. The fluorophores for blue emission have chemical structures like phenyl or fluorene or heterocyclic such as thiophene, pyridine and furan in the polymer backbone. Heterocyclic units such as quinolines, quinoxalines, oxizoles, oxadizoles, benzofunan, thiazoles and thianthrenes are also blue emitting fluorophores when inserted into the polymer backbones. Leung et al. [86] achieved a maximum luminance over 14,070 cd/m^2 for blue light emitting diode based on bis-(2-methyl-8-quinolinato)aluminum hydroxide ($Almq_2OH$) while 14,200 cd/m^2 for bis-(2-methyl-8-quinolinolato)(triphenylsiloxy)aluminum (III) device has also been reported. It is considered that the Al based complexes have high efficiency and are suitable candidates for producing stable blue light emission in small molecular OLEDs.

Enhancement in the performance of OLEDs can be achieved by balanced charge injection and charge transport. The charge transport is related to the drift mobility of charge carriers. Liu et al. [166] reported blue emission from OLED based on mixed host structure. A mixed host structure consists of two different hosts NPB and 9,10-bis(2'-naphthyl)anthracene (BNA) and one dopant 4,4'-bis(2,2-diphenylvinyl)-1,1'-biphenyl (ethylhexyloxy)-1,4-phenylene vinylene (DPVBi) material. They reported significant improvement in device lifetime compared to single host OLEDs. The improvement in the lifetime was attributed to the elimination of heterojunction interface and prevention to formation of fluorescence quenchers. Luminance of 80,370 cd/m^2 at 10 V and luminous efficiency of 1.8 cd/A were reported.

Using a hole blocking layer, holes can be confined in hole transporting or in hole injecting materials and they can be used as emitters. It minimizes the possibility of exciton quenching at the organic/metal interface and enhances the device performance [87]. The devices were fabricated with single and double hole blocking structures, with 4,7-diphenyl,1,10-phenanthroline (BCP), as hole blocking layer. A reduction of 20% in the FWHM for EL spectra and high luminance $\sim$1025 cd/m^2 at 170 mA/cm^2 were reported. A number of organic materials have been used as hole blocking layer like bathocuproine [88], SAlq, 4-biphenyloxolato aluminum (III) bis(2-methyl-8-qoinolinato) 4-phenlyphenolate (BAlq) and tetra(β-naphthyl)silane (TNS). Due to the larger band gap, the problem of large energy barriers for charge injection is more serious for blue light emitting material compared to green and red emitting materials. Generally in device fabrication a large work function ITO (= 4.7 eV) is used as an anode and Ca (= 2.9 eV) as cathode. The work function difference between the two electrodes is less than 2.0 eV, significantly smaller than the band gaps of the blue light emitting materials leading to high injection barriers. Generally hole injection is the main limiting factor in device performance. To fabricate a high performance blue OLED an efficient hole injecting or hole transporting layer is necessary. CuPc has been considered good material for hole injection layer in OLEDs. Its HOMO (4.8 eV) makes the perfect match with the work function of ITO and it does not have the drawback of absorption in the blue spectral region. It even has the capability of filtering out the red and green impure lights. The color purity and stability are the factors, which require improvements in the blue OLEDs. Intense research is going on the development of new blue emitting materials for better performance of OLEDs. The p-i-n structures in OLED devices show high luminance and efficiency at low operating voltages. Some of the blue emitting materials are summarized in Table 4.1. A breakthrough from the limitation of OLEDs was made by a group of researchers from Princeton University who demonstrated very high efficiency by energy transfer from fluorescent host to a phosphorescent guest material. Statistically fluorescent organic materials produce 25% singlet and 75% triplet states in electrical excitation and 100% singlet states in photoexcitation. In fluorescent materials triplet energy states have low emission quantum yield and thus not contribute to electroluminescence. This limits the maximum quantum efficiency for EL to about 25%. But some organomatellic complexes (phosphors) have strong triplet emission quantum yield and provide the possibility of high efficiency EL device by using these materials. The phosphorescent dopants are doped in host materials having slightly higher energy gap. In electro-phosphorescence the energy from both the singlet and triplet states of the fluorescent host is transferred to the triplet state of the phosphorescent guest molecule or the charges are directly trapped by the phosphor. This harvesting of both singlet and triplet states results in 100% internal quantum efficiency. One of the efficient phosphorescent dopant is iridium (III) bis[4,6-di-fluorophenyl)-pyridinato-N,C2']picolinate (FIrpic). OLED devices using this material as dopant have shown external quantum efficiency 10%. Tokito et al. [89] reported a significant improvement in the efficiency of blue OLEDs using phosphorescent iridium complex FIrpic, doped in 4,4'-bis(9-carbazolyl)-2,2'-dimethyl-biphenyl (CDBP). The blue phosphorescent OLED exhibited a maximum external quantum efficiency of 10.4%, corresponding to a current efficiency of 20.4 cd/A. Probably the most recent known phosphorescent blue emitter is iridium (III) bis(4',6'-difluorophenylpyridinato)tetrakis(1-pyrazolyl)borate (FIr6), which shows external quantum efficiency 12% and power efficiency 14 lm/W. Some of the important blue emitting materials, used as host in phosphorescent devices are listed in Table 4.1.

TABLE 4.1
BLUE EMITTING HOST MATERIALS

1. poly(N-vinylcarbazole) (PVK, 466 nm)
2. N,N'-diphenyl-N,N'-bis(3-methylphenyl)-1,1'-biphenyl-4,4'diamine (TPD, 420 nm)
3. 4,4',N,N'-dicarbazole-biphenyl (CBP, 400 nm)
4. 4,4'-bis[N-(1-naphthyl)-N-phenyl-amino]-biphenyl [α-NPD, 430 nm]
5. N,N'-bis-(1-naphthyl)-N,N'-diphenyl-1,1'-biphenyl-4,4'-diamine (NPB, 440 nm)
6. 9,10-bis(2'-naphthyl)anthracene (BNA, 444 nm)
7. bis(2-methyl-8-quinolato) (triphenylsiloxy) aluminum (III) [SAlq, 484 nm]
8. aluminum (III) bis(2-methyl-8-quinolinato) 4-phenylphenolate (BAlq)
9. 9,10-bis(3'5'-diaryl)phenyl anthracene (JBEM, 450 nm)
10. 4,4'-bis(2,2'-diphenylvinyl)-1,1'-biphenyl (DPVBi, 476 nm)
11. bis[2-(2-hydroxyphenyl)benzothiazolate] Zinc [Zn(BTZ)2, 472 nm]

Recently Laskar et al. [90] synthesized some new blue emitting, cyclometallated ligand 2-(4',6-difluorophenyl)-4-methoxypyridine (F_2MeOppyH) based phosphorescent iridium complexes. They fabricated EL devices using the blue Ir(F_2MeOppy)$_2$(acac) complex doped in CBP with different hole blockers BCP and BAlq, and concluded that BCP is a good hole blocker for EL devices. A comparative study of two different host materials CBP and mCP doped with the same blue Ir(F_2MeOppy)$_2$(acac) complex revealed that mCP has better luminance yield and power efficiency than CBP. CBP doesn't have sufficient triplet state energy for effective triplet to triplet energy transfer. A study of the meridional and facial isomers of the same ligand demonstrated that meridional isomer leads to less stable, broader red shifted emission spectrum, and has lower quantum efficiency than the facial counterpart. A dramatic enhancement in the maximum luminance and luminous efficiency from 1400 cd/m^2 and 1.18 cd/A to 5500 cd/m^2 and 3.18 cd/A respectively has been demonstrated by the introduction of perylene as a dopant in the emitting layer. Oligomers, another class of organic semiconductors, are also drawing attention because of their solubility, structure and optical properties similar to polymers and are vacuum evaporable like small molecules. Li et al. [91] reported a fluorine containing arylamine blue emitting oligomer, bis(9,9,9',9'-tetra-n-octyl-2,2'-difluorenyl-7-yl)phenylamine (DFPA) with a maximum luminance of 1800 cd/m^2 and an external quantum efficiency of 1.5%.

4.2.2. WHITE LIGHT EMISSION FROM ORGANIC LEDS

White light LEDs have been fabricated (see Fig. 4.5). The Organic Light Emitting Diode technology has the potential as a direct replacement for LCDs in display applications as well as incandescent lamps for general illumination. All emissive colors are equally important for display applications but a good quality white light is useful for general illumination. The white light source should have good color rendering performance equivalent to that of a black body source at 3000–6000 K temperature and high-energy efficiency. Color rendering index (CRI) is a numerical measurement of how true colors look when illuminated with the light source. CRI can be achieved to its best by broadband spectra distributed throughout the visible region. Color of light is expressed in the CIE (Commission International d'Eclairage) colorimetric system by chromaticity coordinates

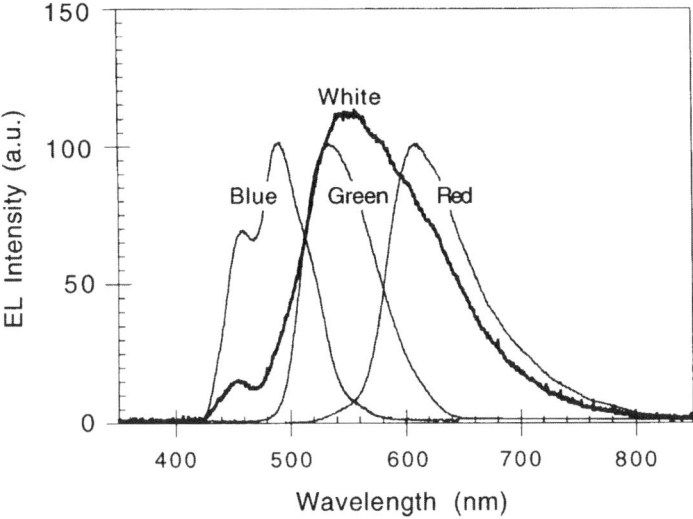

FIG. 4.5. Observed white EL and its blue, green and red components are shown [92].

x and y. A pure white point on the CIE coordinate system is represented as $x = 0.33$ and $y = 0.33$. White OLEDs have typically very broadband emission, which makes them uniquely suitable for application in full color displays, backlight source for liquid crystal displays and general lighting purposes. White light can be produced by mixing light of different colors like blue, green and red in different proportion. Siemens semiconductor group is one of the first companies to commercialize the white inorganic LED [93] developed on the same principle. A blue GaN based LED is used to produce white light on a single chip. The blue light pumps a mixture of suitable phosphors with short pulses. The phosphors emit visible green, yellow or red light and the mixture of these lights produce white light. The efficiency of this device is 20% higher than the incandescent lamp. The working life of the device is 100,000 hours as compared to the life of 50,000 hours of the incandescent lamp.

Since it is very difficult to obtain a single component white light emission from a small molecular and polymeric system, so blending or co-polymerization of more than one organic species are used to achieve white light with good positions on the CIE diagrams. White emission can be obtained by mixing two complementary colors (blue and orange) or three primary colors (red, green and blue) from different sources. Many organic electroluminescent materials have desirable broad emission spectra and the emission color can be tuned by minor changes in their chemical structure. The following techniques are commonly employed to obtain white light emission from organic LEDs:

- Multilayer structure consisting red, green and blue emissive layers.
- Forster/Dexter energy transfer.
- Exciplex/Excimer charge transfer.
- Down conversion by phosphors.
- Microcavity structure.
- White emission by the horizontally/vertically stacked pixels.
- Blending/doping of different emitters into a single layer.

Most of these methods have drawback that the chromaticity changes largely with change in the operating voltage and the fabrication processes are much complex.

4.2.2.1. Multilayer Device Structure

One of the approaches to achieve white light is a multilayer structure where the dopants are segregated into separate emissive layers. Multilayer devices are composed of two or three emitter layers with different emission colors. The emitter layers are prepared by consecutive evaporation or co-evaporation of different emitting materials. To minimize the charge injection barriers and joule heating at the organic/organic interfaces the emissive materials should be chosen such a way that the highest occupied molecular orbital (HOMO) of one material closely matches with that of the different adjacent emissive layers and the lowest unoccupied molecular orbital (LUMO) matches closely with that of the different adjacent emissive materials. The order of the layers is so adjusted that the energy emitted by one layer is not absorbed by the other layer and energy transfer does not take place. The emission of the device depends on the thickness and composition of each layer so they must be precisely controlled to achieve the color balance. The emission intensity can be controlled by varying doping concentrations, adjusting the thickness of layers and inserting blocking layers. The exciton recombination zone can be controlled by inserting a blocking layer that blocks only one type of carrier between the hole transporting layer (HTL) and electron transporting layer (ETL), so that the recombination takes place in two or three different layers resulting emission from each layers. Deshpande et al. [168] achieved white light emission by the sequential energy transfer between different layers. The device was fabricated in the configuration ITO/ α-NPD(200 Å)/α-NPD:DCM2(0.6–8 wt%, 200 Å)/BCP(20–120 Å)/Alq$_3$(300 Å)/Mg:Ag (20:1, 700 Å)/Ag(400 Å). α-NPD was used as hole injection layer, α-NPD:DCM2 layer was used as HTL as well as emitting layer, BCP layer was deposited for the purpose of hole blocking layer, Alq$_3$ was used as green emitting ETL and Mg:Ag followed by 400 Å thick layer of Ag deposited as cathode. A very insensitive spectral response to the current density, with a maximum luminance of 13,500 cd/m^2, maximum external quantum efficiency of 0.5% and an average power efficiency of 0.3 lm/W was reported. It has been found that two or three different colors emitted by different emitters were directly mixed to give white light. This technique requires complex processing and a large amount of wasted organic materials resulting in relatively high fabrication cost. Another approach for white emission from multilayer is multiple quantum well structure, which includes two, or more emissive layers separated by some blocking layers. In this structure both the charge carriers tunnel through the potential barriers and distribute uniformly in different wells. Therefore the matching of the energy levels is not so critical in this system. Excitons are formed and confined in different wells. They do not move to other zones and do not transfer their energy to the next zone. They decay to produce light of different energy in their own wells.

4.2.2.2. Forster/Dexter Energy Transfer

One of the concepts used to produce white light emission in OLEDs is based on the energy transfer mechanism from host molecule to guest molecule. The energy transfer in a host-guest system can take place in two possible ways, either by Forster energy transfer or by Dexter energy transfer. These energy transfers are nonradiative energy transfers. Forster energy transfer involves the dipole–dipole interaction between the donor and acceptor molecules. The energy transfer from host to guest by Forster type energy transfer mechanism is favored by the spectral overlapping of donor emission and acceptor absorption spectra. It allows only singlet-singlet transition at low acceptor concentrations at fast rate $<10^{-9}$ s. This energy transfer has a long-range of about 30–100 Å. Where as Dexter energy transfer involves electron exchange between the host and the guest molecules, having the short-range separation of 6–20 Å. Basically this energy transfer is diffusion of excitons from donor to acceptor. Dexter energy transfer allows both singlet to singlet and triplet to triplet energy transitions. In complete Forster energy transfer the emission from guest will dominant, therefore for a balanced white light emission, incomplete energy transfer is required.

Adjusting the doping concentration of guest in the host material one can get the required color combination to get white light emitting OLEDs. For Forster energy transfer it is necessary that the emission spectrum of host (donor) and the absorption spectrum of the guest (accepter) must overlap. Lot of efforts has been made for achieving of white light emission from small molecules as well as from polymers using Forster/Dexter energy transfer. Mazzeo et al. [81] achieved white emission from a blend of blue emitting TPD host and a red emitting guest, thiophene based pentamer, quinquethiophene-1",1"-oxide (T5oCx), by the incomplete Forster energy transfer from host to guest. The energy transfer was favored by the overlapping of the strong emission spectra of TPD and absorption spectra of the T5oCx. The OLED device having configuration ITO/PEDOT:PSS/TPD:T5oCx/Ca/Al showed a turn on voltage 3.1 V and maximum luminance of 1020 mW/m^2 at 5.4 V, with a balanced white emission, almost independent on the applied voltage.

4.2.2.3. Exciplex/Excimer Charge Transfer

The term "exciplex," strictly used, refers to excited species made by combination of two non-identical moieties, atoms or molecules. Excited complexes that do not fall into this category are known as excimer. In the excimer formation the wave function of excited states extend over the molecules and the molecules are bound together only in the excited state but not in the ground state. The exciplex formation usually takes place at the interface of charge transport layer and the emitting layer. The charge transfer occurs due to the interaction between the excited state of one molecule with the ground state of another molecule, leading to the formation of an electron–hole pair, resulting in radiative decay. The exciplex formation is favored when there is a large difference between the energies of HOMO and LUMO of the emitter and charge transport layer respectively. Due to this large energy difference, the tunneling of charge carriers from transport layer to emitter layer and emitter layer to transport layer will be rare resulting in an accumulation of the charge carriers at the interface. Now the indirect recombination from the LUMO of transport layer to the HOMO of emitter layer is more favored. The energy of the exciplex is always less then the energy of the excited single molecules and its emission is also very broad. A number of research groups have tried white light emission from exciplex formation in small molecules based multilayer devices as well as on polymers based multilayer devices. It has been observed that exciplex formation results in the broadening of electroluminescence spectrum. Recently Singh et al. [95] have obtained white light emission from organic LED based on a simple bilayer structure consisting of TPD and zinc banzothiazole with spectral width of approximately 260 nm. A deconvolution study of PL and EL spectra of the above system revealed that as large as 60% of the broad EL emission originates from multiple exciplex formation.

4.2.2.4. Down Conversion by Phosphors

In several methods used for white emission from organic LEDs, a difficulty of color stability due to differential aging of various species is generally faced. White emission by down conversion phosphors may be an alternative method, which should be far more robust to color shifting as EL intensity decreases with age. In this technology blue emitting OLED is coupled with one or more down conversion phosphor layers, in which one of the phosphor layer contains inorganic light scattering particles. On the backside of the blue OLED, orange and the red phosphors are coated. Some of the portion of the blue light is scattered and goes through the phosphors without down conversion but the rest is converted into orange and red light by the phosphors. As a mixture of all the emitted colors we get white light. Duggal et al. [96] reported white light emission from a blue OLED coupled with a down conversion phosphor system. On the backside of the blue OLED, down converting orange and the red organic phosphor namely perylene orange and perylene red dispersed in poly(methacrylate) (PMMA) were coated, followed by a layer of inorganic light scattering phosphor namely Y(Gd)AG:Ce dispersed in polydimethyl siloxane silicon. The quantum efficiency of photoluminescence dyes in PMMA was found to be >98% and quantum yield of Y(Gd)AG:Ce was 86%. During the operation of the device small portion of the blue light is scattered and goes through the phosphors without down conversion, but the phosphor layers absorb rest of the blue light and emit according to their intrinsic property. The mixing of all the emitted light produces white light. The color tenability can be achieved by changing the concentration and thickness of the phosphor layers.

4.2.2.5. Microcavity Structure

The microcavity is a system consisting of a pair of highly reflecting mirrors having separation of the order of a few micron. The optical microcavity is based on the concept of Fabry–Perot resonant cavity. In the microcavity-structured devices the emitting layer is embedded between two metallic mirrors or a metallic mirror and a partially reflecting bottom mirror composed of distributed Bragg reflector (DBR). It leads to strong modulation of the emission spectrum. In the conventional structures, light is wasted as it leaks in all directions while in case of microcavity structure, light now emerges only from one end of the cavity and the structure is more efficient. Since a microcavity LED is more efficient and uses less current, it lasts longer. Hence microcavity resonator is one of the most effective ways to enhance the luminance and brightness of monochromatic OLEDs. Spectral narrowing and intensity enhancement of spontaneous emission in OLEDs by microcavity has been reported. However it is incapable to white OLEDs because its emission is monochromatic. Dodabalapur et al. [97] demonstrated the controlling of the emission of OLEDs by multimode resonant cavities such that the thickness of the cavity is greater than the single mode cavity devices so that it has several resonant modes within the emission spectrum of the material. By the mixing of the colors from different modes white emission can be achieved.

4.2.2.6. White Emission by the Horizontally/Vertically Stacked Pixels

This technology is similar to the liquid crystal flat panel displays. Here the pixels of the three principal colors are patterned separately and addressed independently. In the horizontally stacked pattern the individual color emitting pixels are deposited either in the form of dots, squares, circles, thin lines or very thin strips. As a result of mixing of these colors any desired range of colors can be produced in the same device. As each color component is addressed individually, the differential color aging can be mitigated by changing the current through the component. Each pixel can be optimized to operate at a minimum operating voltage and for highest efficiency. Also reducing the size of the pixels the lifetime of the device can be controlled effectively.

Burrows et al. [72] have fabricated blue, green and red light emitting diodes in vertically stacked pattern for full color display. Small molecular organic Alq_3 was used as a green emitter, Alq_2Oph as a blue emitter and 3% DCM2 dye doped Alq_3 as a red emitter. α-NPD was used for hole transporting layers. The three LEDs were vertically stacked over each other with transparent electrodes inserted in-between. Each LED was independently addressable. The spectra of the stacked devices are somewhat different from those of individual unstacked devices because of the cavity interaction and coupling effects. 1.0 mm diameter devices operated at 0.1–0.5 mA and 15–25 V. The measured external quantum efficiencies are 0.4% for the green, 0.06% for the blue and 0.05% for the red LED. Unfortunately, the metal layers typically used to connect the individual elements are not very transparent, reducing the resulting brightness of underlying OLEDs in a tandem configuration. Red and blue light had to pass through a Mg:Ag electrode, which had only 50% transmission. External efficiency of 1% is expected from the optimized design of the LEDs in the stacked structure.

4.2.2.7. White Emission by Single Layer Structure

The fabrication process and the device operation of white organic OLEDs from the techniques discussed above are very complex and many parameters need to be optimized for good color rendering and high luminescence efficiency. These complexities can be resolved if a single active layer produces white light emission. White emission from a single layer consisting of a blue emitter doped with different dyes as well as blending two or more polymers, has been reported. But the approach of white emission by two or three colors in a single layer has its own problem that because of different rates of energy transfer between dopants may lead to color imbalance. Some fraction of the highest energy (blue) will readily transfer energy to the green and red emitters and green emitter can transfer energy to the red emitter. If the three emitters are at equal concentrations the red emitter will dominate the spectrum. So the doping ratio must be blue > green > red, at a very careful balanced ratio. A minor shift in the dopant ratio will significantly affect the quality of color. This problem can be solved if a single material is used as an emissive layer and the material has chromophores emitting in the different visible regions. Research is going on the development of white OLEDs based on single molecule as emissive

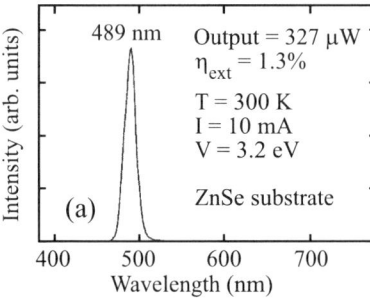
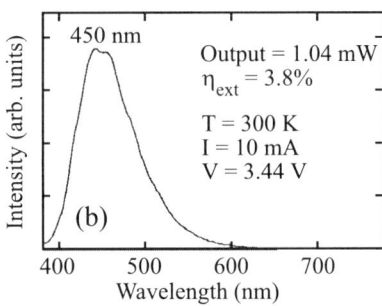

FIG. 4.6. Light output characteristic (a) of the blue ZnSe/ZnCdSe LED fabricated by Eason et al. [99] and (b) of the blue GaN/InGaN LED fabricated by Nakamura et al. [94].

materials. Recently Mazzeo et al. [98] reported a bright single layer white OLED by spin coating a single emitting molecule 3,5 dimethyl 2,6-bis(dimesitylboryl)-dithieno[3,2-b:2',3'-d]thiophene. The white emission was achieved by the superposition of intrinsic blue green light emission of single molecule with a red shifted emission due to formation of cross like dimmers. Bright white electroluminescence was obtained with a maximum luminance of 3800 cd/m^2 at 18 V and the external quantum efficiency of 0.35%.

4.3. Comparison with Other LEDs

Performance of typical II-VI and InGaN blue diodes is shown in Fig. 4.6. The temperature, current and voltage at which the diodes operate are about the same in the two cases. The external quantum efficiency and the output power are about 3 times larger for the InGaN. However spectral characteristics of the ZnSe based diode are superior. Best performance characteristics of the organic and inorganic LEDs are summarized in Table 4.2. The life time of LEDs and lasers are shown in Fig. 4.7.

Since blue, green and red LEDs are now available, white light can be obtained by mixing colors. We have already discussed stacked LEDs which give full color display or

TABLE 4.2
PERFORMANCE CHARACTERISTICS OF INORGANIC AND ORGANIC LEDs [100]

Characteristics	Inorganic LED	Organic LED
Usable brightness	10^7–10^8 cd/m^2	10^2–10^4 cd/m^2
Fabrication	MBE and MOVPE	Spin coating and Vac. Evaporation
Structure	High quality QW	Disordered layer
Spectral quality	FWHM ~20 nm	FWHM ~100 nm
Cost (cm^{-2})	\$$10^2$–$10^3$	\$1–10
Luminous efficiency	Blue: 10 lm/W	Blue: 6 lm/W
Luminous efficiency	Green: 30 lm/W	Yellow–Green: 14 lm/W
Luminous efficiency	Amber: 60 lm/W	Orange–Red: 3 lm/W
Luminous efficiency	Red: 20 lm/W	–
Lifetime	10^5–10^6 h to 80% brightness	6×10^4 h to 60% brightness

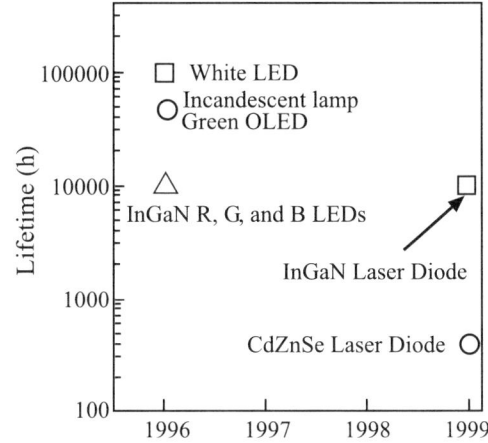

FIG. 4.7. Lifetime of the devices.

white light output [72]. Siemens semiconductor group is one of the first companies to commercialize the white LED [93] developed on a different principle. A blue GaN-based LED is used to produce white light on a single chip. The blue light pumps a suitable phosphor with short pulses. The phosphor emits visible green, yellow or red light. The white light is produced by mixing the colors. The efficiency of the device is 20% higher than the incandescent lamp. The working life of the device is 100,000 hours as compared to the life of 50,000 hours of the incandescent lamp (see Fig. 4.7).

4.4. Organic Solid State Lasers

4.4.1. PHOTOPUMPED LASERS

Photopumped laser action in an organic material was demonstrated in 1967 [101]. First distributed feedback (DFB) and distributed Bragg reflector (DBR) lasers used poly(methyl methacrylate) or gelatin, both doped with rhodamine-6G. Fluorescent organic dyes have been used in spherical resonators. The lasing modes of such resonators are called "whispering gallery" modes. These are optical analogues of the acoustic whispering galleries [101].

The achievement of spectrally narrow polymer laser diode is an important goal for polymer optoelectronic devices. Optically pumped organic lasers work only in short (a few nanoseconds) pulses. The high photoluminescence efficiencies and large cross sections for stimulated emission make the conjugated polymers attractive as the gain material for solid state lasers. Organic materials are cheaper than the most commonly used semiconductor materials in today's lasers, like gallium arsenide. Since organic lasers use organic fluorescent dyes as active medium they can provide emission over the whole visible region. So the small size and availability of organic thin film solid state lasers in the whole visible spectrum makes the possibility to substantially decrease laser cost and an alternative to conventional gas or dye lasers. Optically pumped stimulated emission, gain and lasing action have been observed in many different semiconducting polymers with emission wavelength spanning the entire visible spectral range. The conjugated polymer can also be pumped electrically in diode configuration [102] but so far laser action has not been observed by electrical pumping. The ultraviolet emission from organic materials has been observed in diode configuration but still the realization of organic thin film solid state lasing in the ultraviolet wavelength region below 380 nm is one of the challenges for organic solid state lasers.

4.4.2. SPECTRAL NARROWING

The spectral narrowing occurs only if the refractive index of the polymer film is higher than that of the substrate and if the film is thick enough to support wave guiding. The spectral narrowing results from the amplification of spontaneous emission by stimulated emission as light travels down the waveguide. The spectral narrowing requires two criteria to be fulfilled:

1. The active polymer medium must exhibit Stimulated Emission (SE).
2. The emitted photons should travel a distance greater than the gain length in the excited polymer.

The optical feedback in polymer laser diodes is being provided by microcavities and planer waveguides with distributed feedback. In general a number of explanations have been proposed to explain the spectral narrowing in conjugated polymers, like Amplified Spontaneous Emission (ASE) in asymmetric waveguides, lasing in resonator designs like microcavities and distributed feedback grating structure, cooperative phenomena such as superradiance or superfluorescence and emission from biexcitons or condensed excitons, depending mainly on the excitation geometry. In photopumping of the thin films of conjugated polymers, as the pump intensity is increased the intensity of the spectrum peak grew much more than that of other wavelengths resulting the spectral narrowing.

Light amplification by the stimulated emission process in the absence of resonant feedback can be placed in two categories:

(a) Cooperative process: In this process the dipoles are coupled together through their overlapping radiation fields, and it can be superradiance (SR) or superfluorescence (SF). Superfluorescence (SF) is a phenomenon in which emitters interact with each other over relatively large distance through their overlapping radiation fields and form a coherent state that undergoes cooperative spontaneous emission. For SR the dipoles are initially coherently prepared, while for SF they are initially uncorrelated, although they evolve into a coherent emission after a certain delay time.

(b) Collective process: It is similar to Amplified Spontaneous Emission (ASE). In this phenomenon the spontaneous emission, coming from a distribution of emitters is linearly amplified by the gain medium. Frolov et al. [103] associated the spectral narrowing from DOO-PPV in terms of excitonic cooperative superradiation but McGehee et al. [104] suspected that the possibility of spectral narrowing as a result of superradiance or spontaneous emission from biexcitonic state is wrong. They performed spectral narrowing by photopumping of the sample with a laser stripe of variable length and measured the output light intensity at the end of the stripe. Their results agree with the ASE theory, which predict that the spectra of output light are broad when the pump stripe length is short but become narrow, as the pump stripe length is increased. If superradiance or stimulated emission from a biexcitonic state were responsible for spectral narrowing the emission spectra should not depend on the size of the excited region and the spectral narrowing would have been seen at short excitation length. The dependence of FWHM on the intensity of the pump source at constant stripe length of 2 mm was also described. Organic semiconductor based on Alq$_3$ doped with DCM dye emit in the red region (620–660 nm). Recently organic laser diodes (OLDs) have been fabricated with CBP as the electron transporting layers and doped with perylene. Lasers were also fabricated with C47 and C30 dyes but were not as efficient. Electrons can be easily injected in the CBP from Ag electrodes.

The organic polymers are deposited by spin coating and oligomers by evaporation [106]. The cathodes are deposited by evaporation using shadow masks [106]. Recently patterns of doped polymers were directly deposited by ink-jet printing using a Cannon PJ-1080A printer (see Hebner, 1998 [106]). The polymer used was PVK doped with C6, C47 and nile. Authors have published a beautiful photograph of the ink-jet printed luminescent polymer letters of red nile doped PVK. The thickness of the dots ranged from 400 to 700 Å and the width from 150 to 200 µm.

4.4.3. BLUE LASERS

1996 was a banner year for the organic Lasers. Optically pumped organic laser were reported by several groups [73, and references given therein]. Lasing in current injected organic diodes has not yet been achieved. Optically pumped organic lasers work only in short (a few nanoseconds) pulses. At the present time the organic polymers can not withstand high excitation levels continuously for longer times. With improved sample fabrication techniques the threshold power at which spectral or gain narrowing occurs in the PPV derivatives is now a few tens of $\mu J/cm^2$ [73]. Optically pumped lasers based on dye doped organics and on several polymers have been reported. The emission from a perylene doped CBP film [105] is shown in Fig. 4.8(a). The emission from an electrically driven GaN based laser diode is shown in Fig. 4.8(b) for comparison. The organic laser was pumped with 500 ps pulses of 337 nm light from a nitrogen laser. The repetition rate was 50 Hz. The cavity length was 5 mm. Above the threshold a well defined blue laser beam is clearly observed. The peak wavelength is 485 nm and FWHM is 2 nm. Lasing in CBP doped with 1 to 5% Coumarin 47 was also observed. Lemmer [73] has reported a blue optically pumped laser using LPPP polymer. The LPPP films were 800 nm thick and were fabricated by spin coating. LPPP has two emission bands, one stronger peak at 460 nm and the other vibronic side-band at 490 nm. On increasing the excitation energy to more than 10 nJ, spectral narrowing of the 490 nm emission peak was observed. The absorption band has a low energy tail and the absorption at 460 nm is large. The 460 nm peak does not show the spectral narrowing because of self absorption.

4.5. Quantum Efficiency and Degradation

Internal quantum efficiencies in the inorganic light emitters can approach 100%. Thermal conductivity of the inorganic semiconductors is high. Large currents can be injected in to the diodes. The inorganic semiconductors can be doped to high electron and hole

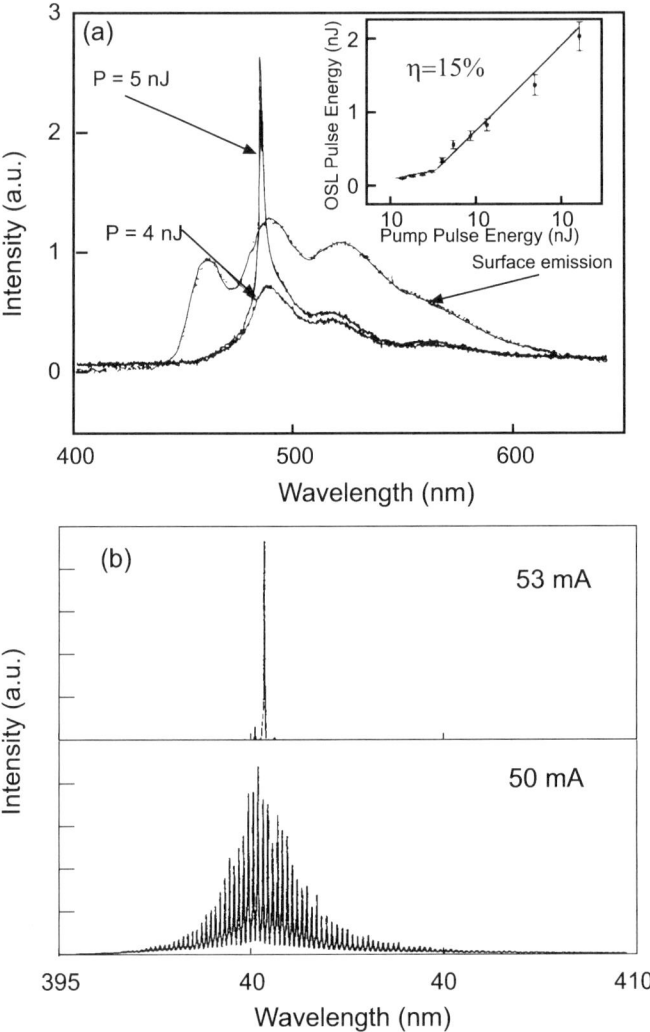

FIG. 4.8. (a) CBP:perylene high quality blue laser emission with very narrow band width [105]. (b) Emission spectra of CW InGaN MQW laser diode with two different operating currents at RT [171].

concentrations. The contact resistance is low for both p- and n-type contacts. Very high quality quantum wells can be fabricated. Therefore the inorganic light emitting devices are inherently more efficient. The excited state in the organic materials consists of triplets and singlets. Only singlets can give rise to EL. Therefore internal quantum efficiency of the EL in organic materials is ⩽25%. In the organic LEDs the degradation increases with the driving current. Hole transporting layers have poor thermal and chemical stability. They can degrade due to poor encapsulation, contamination with impurities emanating from the electrodes and due to heat generated during the working of the device, and also by the injected electrons.

GaN-based LEDs work even when the defect concentration is very high, 10^{10} cm^{-2}. A defect concentration of 10^7 cm^{-2} is fatal for the GaAs-based light emitters. Scientists do not fully understand why these devices work. The maximum lifetime of the ZnSe-semiconductor laser diode is about 400 hours (see Fig. 4.7). The lifetime of the InGaN LEDs and laser diodes is ~10,000 hours. In the case of laser diodes this high value of the lifetime is achieved with the laterally epitaxially overgrown (LEO) layers [107]. The bond strength of III-V semiconductors, including the III-Nitrides is large. Defects are not easily created and they do not multiply because nucleation and propagation energy for the dislocations is large. The II-VI semiconductors have large ionic bonds and are relatively weak. Defects are created during the operation of the devices. They multiply rapidly and migrate into the active layer leading to the failure of the device.

In organic molecular materials the charge carriers are localized on single molecules. In polymers they are localized on single functional groups on the polymer backbone. Low mobilities are not as harmful as one might think. Due to low mobilities the carriers do not sample the large volume before they recombine. Since the charge carriers "see" only the molecule or the functional group over which they reside for long times, they are practically "unaware" of the defects and disorder present in the organic material. The mobilities are sufficient so that the oppositely charged carriers can move and meet each other but not too large to promote non-radiative recombination. Low mobilities also permit very thin (~50 nm) layers of the devices.

We now discuss the inorganic semiconductors. Fabrication of II-VI and III-Nitride devices emitting in the blue region became possible because of the development of strained layer epitaxy [107]. MBE and MOVPE are used for depositing the strained layers of these semiconductors. Most work on II-VI semiconductors has been done on layers grown by MBE. MOVPE has dominated the growth of the III-Nitrides [107]. Generally GaAs substrates are used for the growth of the II-VI semiconductors. Sapphire and SiC substrates have been used extensively for the growth of the III-Nitrides. Luminescence, optical absorption, EPR and X-ray techniques are used to characterize conducting oligomers and polymers and both wide band semiconductor families. In the early days good quality devices could not be fabricated mainly because ZnSe-based semiconductors could not be doped p-type. Successful p-doping of these semiconductors with nitrogen was achieved by using RF-plasma or ECR sources of nitrogen in 1991. Once p-doping was achieved, a junction laser diode using ZnCdSe quantum well (QW) as an active layer could be fabricated. Good ohmic contacts to p-type ZnSe can be made by using a thin ZnSeTe layer. Development of the III-Nitride thin epilayers has been more difficult. A suitable substrate with small lattice mismatch is not available. Bulk GaN crystals of large diameter have not been grown. Quality of layers grown on sapphire and SiC was very poor. Doping GaN with acceptors was also very difficult. These difficulties were solved by breakthrough results obtained in Japan in 1988/1989 [107, and references given therein]. It was found that good quality large area layers of the III-Nitrides can be grown if first a low temperature (~550 °C) 50 nm buffer layer of AlN or GaN is grown by MOVPE on a sapphire substrate. After the growth of the buffer layer, the temperature is raised to ~1050 °C for the growth of the main epilayer of the III-Nitride. As grown Mg doped GaN is semi-insulating. It was also discovered that Mg doped GaN becomes p-type highly conducting if irradiated with low energy electron beams. Annealing at high temperatures also activates Mg acceptors. Making low resistance ohmic contact to p-type GaN has not yet been accomplished. Blue and green InGaN active layer LEDs were commercialized in the mid-1990s.

4.6. Stability

4.6.1. DEGRADATION OF THE POLYMER

Degradation of the LEDs can occur due to several factors. Some of these factors are: crystallization of the organic layer, electrochemical reactions at the electrode/organic interface, migration of ionic species, oxidation of the polymer, and electrochemical and electrooptical reactions.

The lifetime of an LED is defined as the time during which the emission of the device decreases by 50%. For commercial application of the organic LEDs the shelf life should be several years. The operational lifetime should also be sufficiently long, i.e. between 100 and 10,000 hours. The minimum lifetime required depends on the application. The intrinsic lifetime of a device under ambient conditions is short due to degradation by oxygen and water vapors. Degradation depends on the morphology and threshold voltage for light emission. These properties depend on the nature of the solvent as discussed below.

Liu et al. [108, and references given therein] have studied the effect of nature of the solvent, polymer concentration in the solvent and of the spin speed on the electrical and optical properties of poly(2-methoxy-5-(2'-ethyl-hexyloxy)-1,4-phenylene) vinylene (MEEH-PPV) based light emitting diodes (LEDs). The cathode was Ca or Al metal and the anode was ITO/PEDOT (3,4-polyethylenedioxythiophene-polystyrenesulfonate). Devices were fabricated in two solvents (1) the so-called DCB films and devices from a 0.8 wt% (8 mg/ml) solution in dichlorobenzene and (2) or the THF devices from a 0.6 wt% (6 mg/ml) in tetrahydrofuran. The polymer thin film spun directly from the THF solution was usually not uniform. Therefore the THF solution was spun under an environment in saturated THF vapors. The thickness of both the DCB films and the THF films was about 130 nm. The experimental I–V and B–V curves showed significant differences in the behavior of the two films. The DCB films showed a higher injection current. The DCB films also had a higher mobility and lower electrical resistance. The DCB devices had a better contact with the anode and therefore lower barrier for hole injection. The voltage for light emission was smaller for the DCB devices; 1.75 V for a DCB device versus 1.94 for a THF device. Because of the higher voltage needed for light emission, the THF devices are likely to degrade earlier.

The properties of anode/polymer contact depend on the morphology of the polymer film. Changing solvents and fabrication conditions can control the morphology. In the case of non-aromatic solvents there is a time lag in the current injection voltage (i.e. electron injection voltage) and light emitting voltage (which is equal to hole injection voltage). Experiments show that the solvent induced changes in morphology are important in determining the device performance and presumably the degradation.

Early studies of the degradation and failure of ITO/MEH-PPV/Ca light emitting diodes were made by Scott et al. [178]. The authors found that the degradation occurs by two

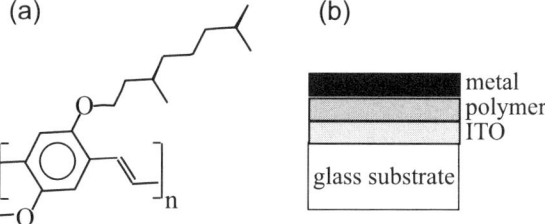

FIG. 4.9. Chemical structure of poly(dialkoxy-p-phenylene vinylene) [109]. (b) Schematic layer structure of the device.

mechanisms. The first mechanism results from oxidation of the MEH-PPV polymer itself, probably the oxygen atoms come from the ITO anode. The oxidation decreases the luminous efficiency and increases the electrical impedance. The efficiency decreases on both accounts. The second mechanism involves the formation of hot spots at the cathode, similar to the black spots observed by subsequent workers and discussed below. As the hot spots increase in size the active area of the device is progressively eroded until the device fails.

Berntsen et al. [109] measured the stability and lifetime of the organic LEDs. The active layer of the device consisted of a single layer of poly(dialkoxy-p-phenylenevinylene). The chemical structure of the polymer is shown in Fig. 4.9(a). A schematic layer structure of the LED is shown in Fig. 4.9(b). The operational lifetimes in nitrogen ambient in a glove box were a few hours. The lifetime improved on treating the ITO with UV/O_3. Typically the lifetime was 300–400 h for such devices. In the light and in the air photooxidation of the PPV occurs. The vinyl-bonds break and carbonyl groups are formed. The carbonyl groups are observed in the Fourier Transform Infrared Spectroscopy (FTIR). The conjugation of the polymer is interrupted and the conjugation length of the polymer is reduced. The optical absorption and the photoluminescence intensities decrease and their peaks shift to shorter wavelength. Experiments showed that this type of degradation was more severe in air than in water vapor or vacuum. The degradation was not sensitive to the incident power density over wide range of the power density values. Optical measurements also showed that the degradation under illumination did not occur if the ambient is nitrogen. X-ray photoelectron spectroscopy (XPS) showed that the thickness of the oxide layer at the interface of the cathode and the active layer did not increase and the devices did not degrade during operational testing. Therefore we conclude that for a single layer device, the lifetime is not limited by the degradation of the polymer at the interface.

4.6.2. THE CATHODE AND THE BLACK SPOTS

The cathode degrades by the formation of black spots if the device is operated under ordinary ambient conditions [109]. The black spots are circular in shape and do not emit any light. With aging the average size of the spots increases but their number remains constant as shown in Fig. 4.10(a). SEM measurements showed that there was a pinhole at the center of each spot. The black spots are cause by the pinholes in the metal. The number of the black spots decreased as the thickness of the metal layer was increased as shown

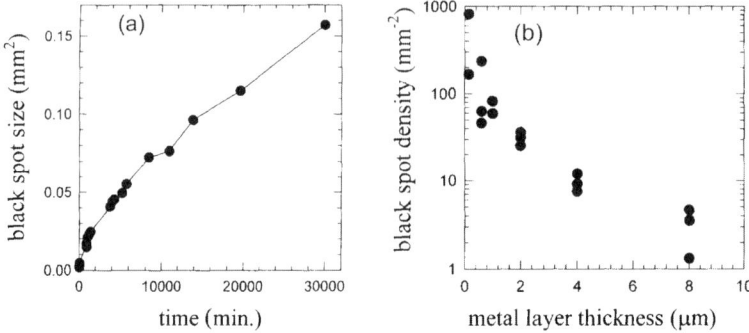

FIG. 4.10. (a) The average black spot size as a function of storage time in air for an unencapsulated single layer device with a 4 μm thick metal cathode. (b) The black spot density versus the thickness of the cathode layer [109].

in Fig. 4.10(b). Subsequently black spots have been observed by several other groups [110,111, and references given therein]. Schaer et al. [110] fabricated the organic LEDs on ITO coated glass substrate. The layer structure of the device is as follows: 10 nm layer of copper-phthalocyanine (CuPc) evaporated on ITO, 40 nm thick hole transmitting layer 4,4'-bis(N-(1-naphthyl)-N-phenyl-amino)biphenyl (α-NPD), and 25 nm of Alq$_3$ doped with rubrene. The cathode was a 0.9 nm thick LiF layer covered with 75 nm thick Al layer. Following fabrication the LEDs were tested inside a glove box. Experimental results of Schaer et al. [110] are shown on the left in Fig. 4.11. Fig. 4.11(a) is the optical image of an LED working at a luminance of 100 cd/m^2 under water vapor atmosphere. Non-emitting dark spots can be seen clearly. If the device is operated at a constant brightness, the growth of the spot diameter Ds is given by,

$$Ds(t) = A_0 \exp(kt), \tag{4.1}$$

where A_0 is the initial dark spot diameter and $k = 2 \times 10^{-4}$ s^{-1} is the growth rate. Fig. 4.11(b) shows the SEM picture of the black spots. Fig. 4.11(c) (on the left side) shows that as the diameters of the spots grow, the driving current decreases. The spots hinder the injection of the carriers into the polymer. Schaer et al. [110] also showed that water is a thousand times more destructive degrading agent than oxygen at room temperature.

Czerw et al. [111] investigated the failure modes of an LED with MEH-PPV emissive layer. The layers were grown on B doped Si substrate and the layer sequence was Ag, Si mono-oxide, PEDOT:PSS (polyethylene dioxy-thiophene-polystyrenesulfonate), MEH-PPV, Alq$_3$ and LiF/Al. The black spots observed in a cross-sectional scanning microscope are shown on the right side in Fig. 4.11. The size of the spots varied from hundreds of nanometers to several microns. Energy dispersive X-ray analysis showed that the black spots mainly consisted of carbon. At the location of the spots the polymer has been carbonized and has extruded through the cathode. The regions surrounding the spots are left with no polymer. Delamination of the cathode and its separation from the polymer was also observed.

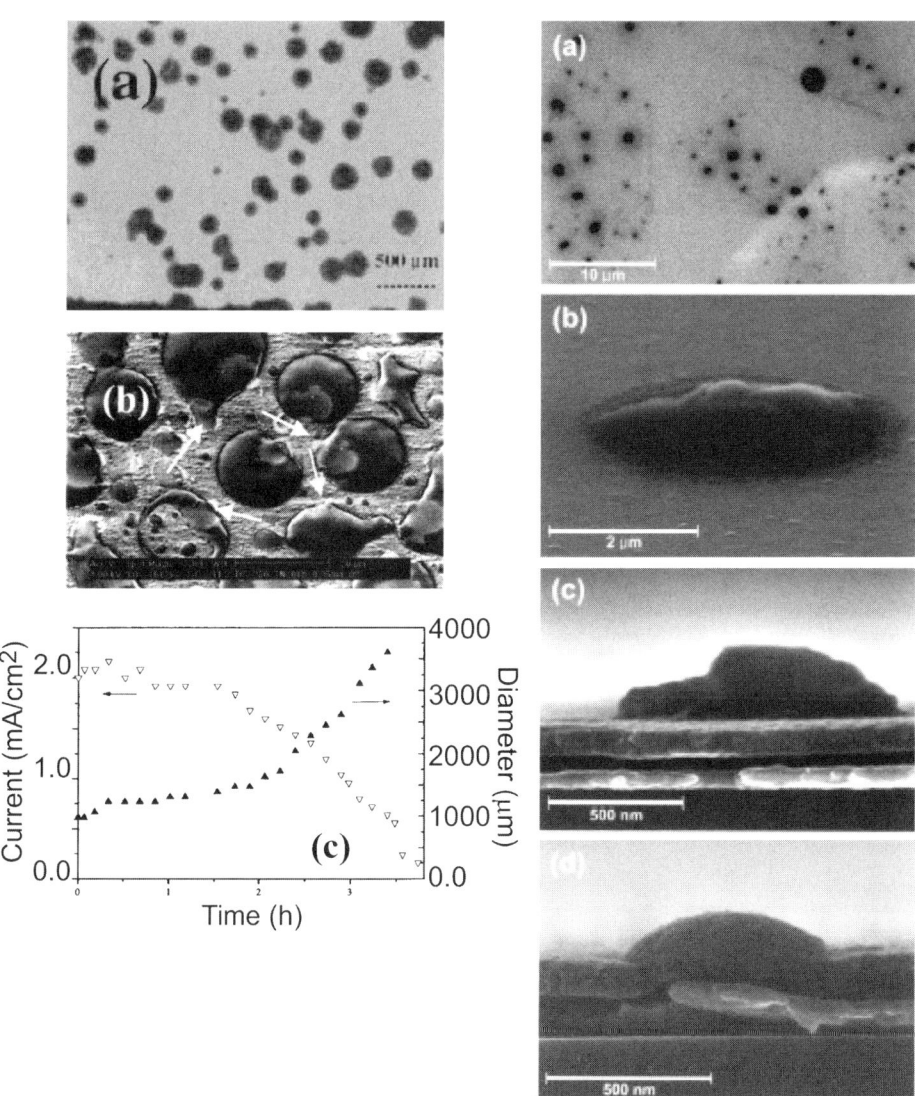

FIG. 4.11. Left: (a) Optical microscope image of an OLED working at a luminance of 100 cd/m^2 under water vapor atmosphere. Non-emitting dark spots can be seen clearly. (b) SEM image of the bubbles formed on the aluminum cathode in the dark spot area. (c) Correlation between dark spot growths (taken from the increase in diameter) and total current density [110]. Right: (a) Shown here is the random pattern of carbonized areas on the surface of the cathode after operation, shown in wide field. (b) At higher resolution, the structure of one of these areas becomes more apparent. (c) and (d) show nanoscale views of carbonized areas with the extrusion of the polymer through the cathode and the resulting void underneath [111].

4.6.3. DEGRADATION OF THE ANODE

Chemical cleaning of ITO and its treatment with UV/O_3 or O plasma improves the performance of the LED considerably. The device works at lower voltages and for longer times. Measurements show that this treatment increases the work function of the anode considerably. Treatment with H_2 did not increase the work function. Therefore it is concluded that the treatment increases the oxygen content of the ITO surface. Oxygen is electronegative and can form a negatively charged surface layer. This can cause a depletion of electrons just below the surface. The resulting band bending can increase the work function. However this oxygen layer is not very stable. The stability of the anode can be improved by applying a PEDOT layer on ITO before coating it with the emissive polymer layer [109]. The PEDOT layer increases the operational lifetime enormously. Tests in a glove box gave lifetime up to 5000 hours. Berntsen et al. [109] developed a technique of encapsulation such that oxygen and water vapors do not reach the active layer. The performance of the encapsulated LED with PEDOT/ITO anode is shown in Fig. 4.12. Nine devices were tested. The operational lifetime of 7 of these devices was more than 3000 h. The devices were also tested at high temperatures. At 70 °C and 50% humidity the life time was more than 500 h.

Jeong et al. [112] have discussed again recently the problem of cathode degradation in LEDs. Recent studies show that the dark spots on the cathode have dome like bubble structure filled with gases, mostly oxygen. The authors found that if Al cathode is prepared by Ion Beam Assisted Deposition (IBAD), it is more dense and suppress the formation of the black spots. However the energetic ions degrade the organic material and performance if the device degrades. In normal LED processing Al cathode is deposited by thermal evaporation. Thermal evaporation avoids damage to the organic material caused by energetic ions in IBAD. By inserting a thin Al buffer layer between the IBAD Al and organic material the damage to the organic material was suppressed. Jeong et al. [112] were able to fabricate OLEDs with good performance and longer life.

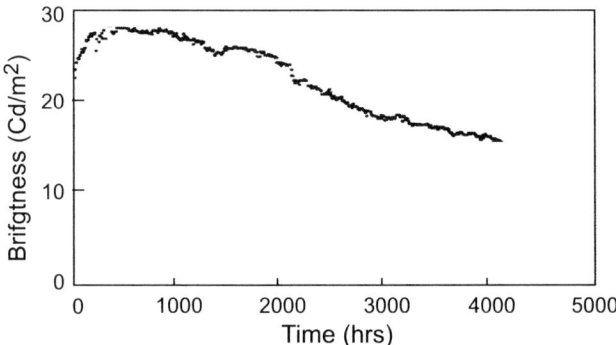

FIG. 4.12. Light output of an encapsulated 8 cm^2 polymer LED with an ITO/PEDOT anode. The measurement has been performed in a climate chamber at 20 °C and 50% relative humidity [109].

4.7. Soluble New 5-coordinated Al-Complexes

The most important work in the luminescent organic materials was the discovery of luminescence in tri (8-hydroxyquinolinato) aluminum (Alq3) [66]. Lots of 8-hydroxyquinoline derivatives have been synthesized for the purpose of color tuning and better device performance. In all cases of 8-hydroxyquinoline derivatives the emission (PL and EL) have been attributed to the π–π^* transitions in 8-hydroxyquinoline ligand. Both blue and red shifts in the emission can be obtained by introduction of electron donating substituent on the pridyl ring or on the phenoxide ring [113]. Alq3 films are usually prepared by vacuum deposition, which require costly coating units. Alq3 has been reported as a pure green emitter with a broad emission peak at 530 nm. Pure yellow and red emission can be obtained by doping Alq3 with a small amount of suitable fluorescent dyes [113].

Misra et al. [113] have reported the synthesis and optical/electrical properties of new 5-coordinated Al-complexes designed as Alq(1) and Alq(2). The complexes are vacuum evaporable as well as soluble in many organic solvents. EL peaks of these new complexes emit in the range 522–523 nm, which is nearly 8 nm blue shifted compared to that of Alq3. The chemical structures of the complexes were determined with the help of the Hydrogen Nuclear Magnetic Resonance (HNMR) and Fourier Transform Infrared (FTIR) spectroscopy techniques. The structure of these complexes is shown in Fig. 4.13.

The thermal stability of the complexes was determined by thermo gravimetric analysis (TGA) with a Mettler TA 3000 System at a scan rate of $10\,°C/min$ under nitrogen atmosphere. The complexes were found quite stable up to $430\,°C$. The maximum weight loss occurred at about $450\,°C$.

UV-vis absorption spectra of Alq(1) and Alq(2) were studied in the solution and in thin films. No difference was observed in the main absorption peaks in solution and thin films. The absorption and PL measured in the solution and in thin films of the complexes are shown in Fig. 4.14. The synthesized complexes showed strong green emission in the range 522–523 nm. The optical band gap of the materials were determined from absorption edge and using the Tauc relation, which relates the absorbance A, with the band gap as

$$Ah\nu = (h\nu - E_g)^n \qquad (4.2)$$

where n is $1/2$ for direct band gap materials and 2 for indirect band gap materials, E_g is the band gap and $h\nu$ is the photon energy. The value of the optical band gap was obtained by extrapolation of the straight-line portion of the Absorbance^2 versus $h\nu$ plot, to $A = 0$. The calculated band gaps of the complexes were nearly equal at a value 2.8 eV.

For Alq(1) and Alq(2) the EL spectrum peak appeared at 522 nm and 523 nm respectively. The values are nearly 8 nm blue shifted compared to that of Alq3 (531 nm). All the EL spectra of the devices are almost identical to the PL spectra of respective aluminum complexes.

The OLEDs were fabricated on the pre-patterned, pre-cleaned indium tin oxide (ITO) coated glass substrates. The substrates were patterned using standard photolithography technique and then cleaned with soap solution followed by boiling in trichloroethylene and isopropyl alcohol. The films were finally dried under vacuum. After cleaning the

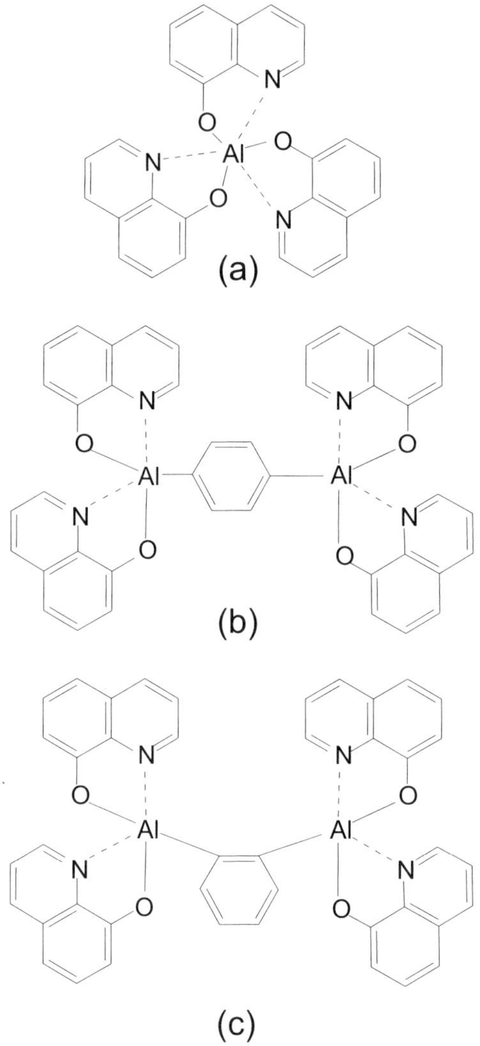

FIG. 4.13. Molecular structure of (a) Alq$_3$, (b) Alq(1) and (c) Alq(2) [113].

substrates were treated with atmospheric plasma for five minutes. The devices were fabricated in the double layer configuration ITO/TPD(25 nm)/Alq(1)(35 nm)/LiF(0.5 nm)/Al(150 nm) (device 1) and ITO/TPD(25 nm)/Alq(2)(35 nm)/LiF(0.5 nm)/Al(150 nm) (device 2). Alq$_3$ device was also fabricated in the same configuration ITO/TPD(25 nm)/Alq$_3$(35 nm)/LiF(0.5 nm)/Al(150 nm) (device 3). Device 3 was used as a reference device. In all devices the hole-transport layer (TPD, 25 nm) was evaporated onto the cleaned ITO substrates to facilitate better injection of holes into the active layer (aluminum complex layer). Subsequently a 35 nm layer of an aluminum complex was deposited on TPD. After a thin buffer layer of LiF (0.5 nm) a 150 nm thick Al cathode was deposited in the same vacuum condition. Thickness was measured with a HINDHIVAC quartz crystal

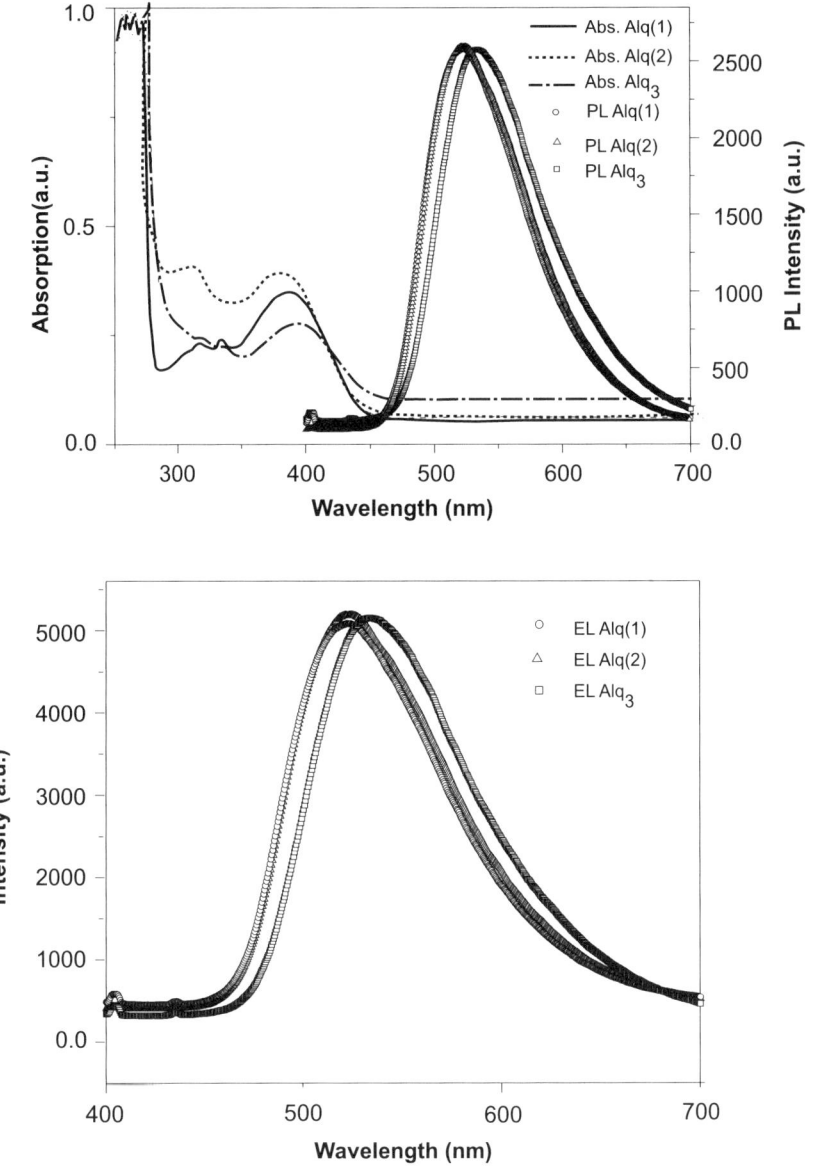

FIG. 4.14. Upper figure: Normalized UV-Visible and PL spectra of Alq(1), Alq(2) and Alq$_3$ aluminum complexes. Lower figure: Normalized EL spectra of Alq(1), Alq(2) and Alq$_3$ aluminum complexes [113].

DTM-101 thickness monitor. After fabrication the device was transferred to a glove box, so that there was a brief exposure of the device to air. The device was hermetically sealed in the glove box under dry nitrogen atmosphere and then was taken out. Subsequently all the electrical measurements were done in the ambient atmosphere. OLEDs were also fabricated by the spin casting of Alq(1) and Alq(2) from the chloroform solution. The

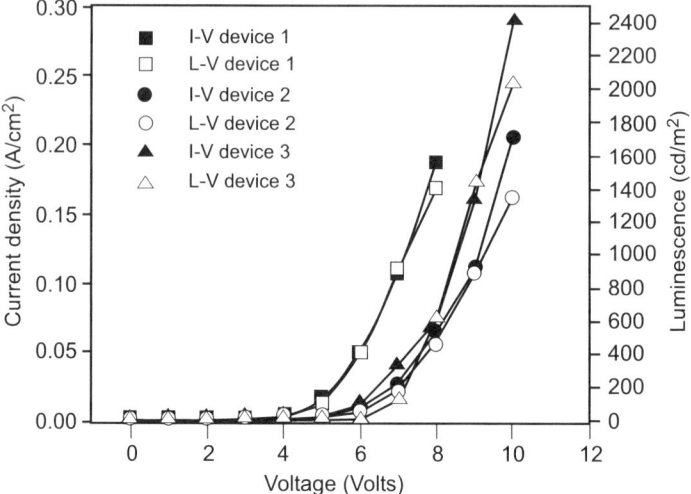

FIG. 4.15. I–V–L characteristics of devices in ITO/TPD/aluminum complex/LiF/Al configuration [113].

EL of solution-processed devices was similar to that fabricated by vacuum evaporation of the complexes but the operating voltage was quite large. The I–V–L characteristics of the devices were measured using computer interfaced KEITHLEY 2400 Source Meter and LMT luminescence meter. The I–V–L characteristics of device 1, device 2 and device 3 are shown in Fig. 4.15. The current and light output increased with the forward bias voltage. The turn on voltages of device 1, device 2 and device 3 were 4, 6, and 5.5 V respectively. For device 1 the luminance of 1400 cd/m^2 was achieved at the applied voltage of 8 V. For device 2 and device 3 the luminance of 1342 and 2040 cd/m^2 was achieved at the applied voltage of 10 V. The luminous efficiency was 0.74, 0.65 and 0.70 cd/A for device 1, device 2 and device 3 respectively. The emission was uniform over the entire active area of the devices. EL for all devices was studied at different voltages from 4 to 10 V and was found to be nearly independent of the applied voltage and CIE coordinates stayed at nearly same position.

4.8. Summary and Conclusions

Scientists and engineers have made considerable advances in the field of organic LEDs during the last 10 years. Several blue, green and white solid state light sources have matured. Many of the devices have been commercialized. More work is still required in some areas. Long term reliability of II-VI devices needs improvement. Better contacts to p-GaN are urgently required. Life time of optically pumped organic lasers is still short. It is not clear whether scientists will be able to achieve *electrically pumped* organic lasers.

CHAPTER 5
SOLAR CELLS

5.1. Introduction

Solar cells (also known as photovoltaic (PV) devices) have become an important source of renewable clean energy. The world annual production of electricity by Si solar cells is now several tens of megawatts. However the cost of these cells is high. They are not able to compete with the conventional power plants. The search for cheaper PV cells has been on for a long time [114,7]. Early work on conducting polymer solar cells was done with the hope that the cost of the cells will come down. However the conversion efficiency η of the early cells was only a few hundredths of a percent, always smaller than 0.1%. The main difficulty arises due to the fact that the excitons generated by the incident light in the conducting polymers have a large binding energy. They do not dissociate at room temperature. The probability of their recombining is much higher than the probability of their dissociation. This property is useful for light emitting devices and explains why the work on LEDs has been so successful (see for example [115,92,105]). Until recently the prospects of useful polymer PV cells were not good. However two important papers were published in 1992 [116,117]. These papers showed that Buckminsterfullerene C_{60}, mixed with a conducting polymer, is very effective in dissociating the excitons in the polymer. PV devices were made without the use of exciton breaking agents in 1994 [118,15]. These devices continued to show poor performance. More recently PV devices have been fabricated using other exciton breaking agents [34,119–123,35,124]. These papers have revived the hope that cheap energy conversion using polymer cells may become a reality. Large area flexible photodetectors useful for many applications can be fabricated. Dye sensitized high efficiency solar cells have also been fabricated. We discuss the work on these PV devices and identify the areas where more research is urgently required.

The technology used in manufacturing organic devices is simple and does not use expensive equipment. Organic solar cells are cost effective and can be fabricated on large area and flexible substrate. Recently extensive work has been done on conducting organic solar cells and dye sensitized solar cells. The conducting organic semiconductors are now challenging the dominance of the photovoltaic field by Si and other inorganic semiconductors [125,126].

5.2. Solar Cells

5.2.1. SINGLE AND BILAYER SOLAR CELLS

A single layer solar cell is the simplest solar cell. It is essentially a Schottky diode with a polymer layer sandwiched between two metal electrodes. Several groups have measured the $I-V$ curves of dark and illuminated polymer Schottky diodes under both the forward and reverse bias [127, and references given therein]. We discuss the dark and illuminated characteristics of the ITO/PPV-MEH/Ca diodes. The reverse current in the dark is small and the diode shows reasonable rectifying characteristics. The currents become 0 at voltage $V_{oc} = 1.4$ eV. This value agrees approximately with the difference in the work functions of ITO and Ca. In fact in better quality diodes the observed value of $V_{oc} = 1.6$ eV, which is in excellent agreement with the difference in the work function values (see also [166]). Wei et al. [120] also fabricated diodes with Al and Cu cathodes. In both cases the observed value of V_{oc} was equal to the difference of the work functions of ITO and the cathode metal used. Since the difference in the work functions of ITO and Cu is small, the dark and illuminated characteristics are nearly symmetrical with respect to $V = 0$ axis and $V_{oc} \approx 0$. The conversion efficiency of the single layer cells was low.

Several authors have investigated the bilayer solar cells in an effort to improve the efficiency of the cells. In these solar cells the active layer consists of two (donor and acceptor) layers with a planar junction. We discuss here recent work of Breeze et al. [128]. These authors studied optical absorption of PBI, M3EHPPV and MgPc (shown in Fig. 5.1) and characteristics of bilayer solar cells fabricated using these organics. The authors studied ITO/M3EH-PPV/PBI/Au solar cells in detail. The short circuit current was 1.96 mA/cm^2, the open circuit voltage was 0.63 V, and the fill factor was 46%. They also studied the cell fabricated by reversing the order of the polymer and perylene layers but keeping the same configuration of the electrodes. The performance deteriorated considerably. Though the value of the open circuit voltage remained about the same, the short circuit current decreased by a factor more than 3. The open circuit voltage is more than the difference of the work functions of the electrodes. This shows that unlike the single layer devices discussed in the beginning of this section, the open circuit voltage is not determined only by the work function values of the electrodes. Photocurrent action spectra for ITO/PBI(20 nm)/M3EH-PPV(35 nm)/Au (circles) and ITO/M3EH-PPV(44 nm)/PBI(24 nm)/Au (squares) are shown in Fig. 5.1(a). Both the polymer and

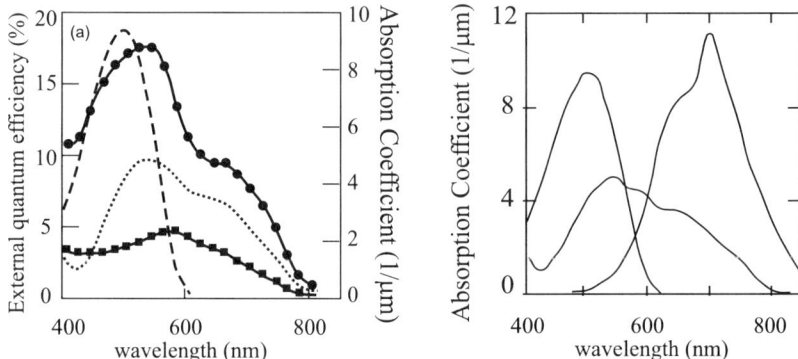

FIG. 5.1. (a) Photocurrent action spectra for ITO/PBI(20 nm)/M3EH-PPV(35 nm)/Au (circles) and ITO/M3EH-PPV(44 nm)/PBI(24 nm)/Au (squares) on the left-hand side axis. The absorption spectra of M3EH-PPV (dashed line) and PBI (dotted line) are shown on the right-hand side axis for comparison. (b) Absorption spectra of M3EH-PPV (left curve), PBI (middle curve) and MgPc (right curve) [128].

TABLE 5.1
Values of Open Circuit Voltage V_{oc} (V), Short Circuit Current I_{sc} (mA/cm^2), Collection Efficiency η_c (%), Conversion Efficiency η (%) and Fill Factor (FF) of the Polymer Solar Cells and Photodiodes

Polymer	Contact	V_{oc}	I_{sc}	η_c	η	FF	Ref.
PPEI	Ag	0.23	–	–	–	–	[123]
PPEI	Ag + TPD	0.22	–	–	–	–	[123]
PBI/MgPc	Au	–	–	–	0.17	–	[128]
PBI/M3EH-PPV	Au	0.63	0.65	–	–	–	[128]
M3EH-PPV/PBI	Au	0.63	1.96	–	0.71	0.46	[128]
MEH-PPV:CN-PPV	Ca or Al	1.2	0.08	5	0.9	–	[34]
MEH-PPV:C$_{60}$	Ca	0.43	1.8	26	2.5	0.65	[121]
MEH-PPV:CN-PPV	Ca or Al	1.2	0.08	29	2.9	–	[119]
MDMO-PPV:PCBM	Al + LiF	0.834	–	–	3.2	0.632	[124]
P3HT:PCBM	Al + LiF	–	8.5	70*	3.5	–	[129]

In all cases included in the table, the contact layer on one side was ITO. The contact layer on the other side is also given in the table.
* External quantum efficiency.

perylene PBI contribute to the photocurrent. Fig. 5.1(b) shows the absorption of M3EH-PPV, PBI and MgPc. From the analysis of the spectra the authors concluded that the exciton dissociation occurs only in regions very close to the interface. Note that the photocurrent peak of PBI(20 nm)/M3EH-PPV(35 nm) occurs at the absorption edge rather than at the maximum. The performance of the solar cells is given in Table 5.1. As shown in Table 5.1, the performance of solar cell fabricated using small molecule PBI and MgPc was very poor.

It is seen from Table 5.1 that the values of the conversion efficiency in bilayer solar cells also is quite low. As mentioned in the introduction it is difficult to dissociate excitons in the conducting polymers. The Donor/Acceptor (D/A) junction between the polymer and the fullerene is rectifying and can be used for designing photovoltaic cells or photodetectors. In this bilayer cell also the conversion efficiency is low. The cause of the low efficiency is that the charge separation occurs only at the D/A interface that results low collection efficiency. The diffusion length of the exciton is a factor 10, lower than the typical penetration depth of the photon.

5.2.2. Interpenetrating Network of Donor–Acceptor Organics. Bulk Heterojunction Solar Cells

The situation improved considerably when the active layer was formed by a network of two (donor and acceptor) interpenetrating polymers or small molecules [34,119,121–123,35]. For example Buckminsterfullerene C$_{60}$ mixed with MEH-PPV is very effective in dissociating the exciton created by the incident light. C$_{60}$ acts as an acceptor and the polymer, as a donor. The transfer of electron from MEH-PPV to fullerene occurs because fullerene has a larger electron affinity. The hole is left at the MEH-PPV because it has small ionization potential. The exciton dissociation results in quenching of the PL by factor up to 10^4 and in increasing the photoconductivity considerably. The transfer rate

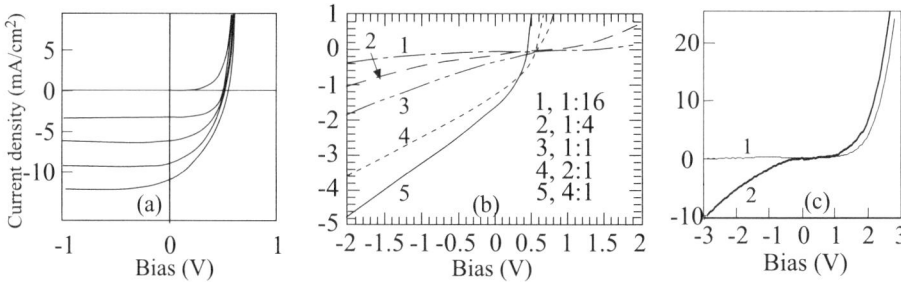

FIG. 5.2. $I-V$ characteristics of (a) an illuminated Si photodiode [120] and (b) of an ITO/C_{60}:MEH-PPV/Ca solar cell [121]. Curve numbers and corresponding values of the concentration ratio R of fullerene:MEH-PPV are given in figure (b). (c) Dark (curve 1) and illuminated (curve 2) $I-V$ characteristics of an ITO/C_{60}:PS:Ooct-OPV5/Al device with an active area of 0.24 cm^2 [127]. Note the difference in the electric field dependence of the illuminated reverse bias current of the organic and Si diodes.

of electrons from the polymer to the C_{60} is 1000 times faster than the decay rate of the photoexcitations. The efficiency of transfer (electron/photon) is close to unity. If the active layer of a solar cell consists of such a network, the cell is known as a Bulk Heterojunction Solar Cell (BHSC). The effect of composition of the BHSCs on their characteristics is shown in Fig. 5.2. Fig. 5.2(b) shows that the performance of the cell improves rapidly as the concentration of the C_{60} increases.

We show the measured $J-V$ curves of a Si solar cell also in Fig. 5.2(a). As expected, the reverse bias current of the illuminated Si solar cell is independent of the applied voltage. The behavior of the organic solar cells shown in Figs. 5.2(b) and 5.2(c) is strikingly different. The current increases strongly with the applied reverse bias. Two models have been suggested to interpret these observations. In the first model it is suggested that the effective mobility increases with the applied bias, an effect that is absent in the Si solar cell. However since the free carrier density is very large under illumination and all the traps are likely to be filled, we do not think that the mobility can be a strong function of the applied bias [127]. In the second model, the applied bias increases the dissociation of excitons and therefore the number of free carriers. The increase of dissociation of the excitons by the electric field should suppress the PL intensity. Kersting has confirmed this experimentally as shown in Fig. 5.3.

Yu and Heeger [34] and Halls et al. [122] used the two polymers, CN-PPV and MEH-PPV, to form the interpenetrating network of the active layer. Both polymers are soluble in chloroform and samples can be easily prepared by spin coating. The PPVs are good hole transporting materials and are widely used in the electronic devices. By adding the CN side groups, the ionization potential and electron affinity increase by about 0.5 eV. This improves the electron injection and transport properties of the polymer. Electron affinity and turn-on voltage of CN-PPV is larger than the corresponding quantities of the MEH-PPV. Electrons are transferred from MEH-PPV to CN-PPV and holes are transferred from CN-PPV to MEH-PPV.

The mixture of the two polymers tends to segregate due to low entropy of mixing. The phase segregation has been confirmed by TEM, STEM and PEELS (parallel electron-energy-loss spectra). Care must be taken to prepare the samples of acceptable structure.

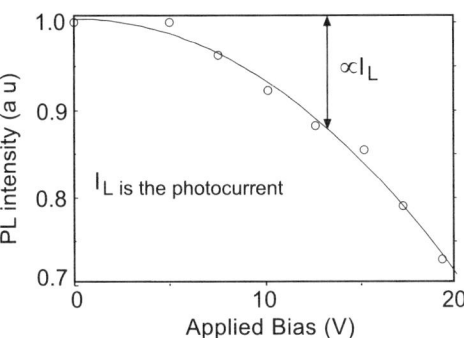

FIG. 5.3. The observed decrease of PL intensity by the applied electric field. The decrease occurs because of the dissociation of excitons by the electric field. We have reported Kersting's data [29] in a different format. The photocurrent I_1 increases rapidly.

The absorption spectra [34] of the MEH-PPV, CN-PPV and of the mixed layer consisting of penetrating network of the two polymers are shown in Fig. 5.4. The absorption spectra show that there are no peaks below the absorption edge. There are no states in the bandgap and there is no charge transfer from the donor polymer to the acceptor polymer. The difference in the electro-negativity is not sufficiently large for the charge transfer doping. However it is seen from Fig. 5.4 that the PL of the mixed layer is quenched, confirming that the charge transfer between MEH-PPV (donor) and CN-PPV (acceptor) has taken place under the action of the incident light. Strong quenching of the PL has also been observed by Halls et al. [122] and by Yoshino et al. [35]. The photoresponse increases considerably as a result of this quenching. In these experiments the PL is quenched by a factor from 15 to 25. These results suggest that the blend of these two PPVs is suitable for fabricating BHSCs.

The dark and the photocurrents of the Ca/MEH-PPV:CN-PPV/ITO solar cells are shown in Figs. 5.5(a) and 5.5(b). The collection efficiency (electron/photon) was 5% and conversion efficiency was 0.9%. The efficiencies are 20 times larger than the MEH-PPV cell and 100 times larger than the CN-PPV cell. The photosensitivity increases considerably under reverse bias, to ~ 0.3 A/W and 80% electron/photon at -10 V. The open

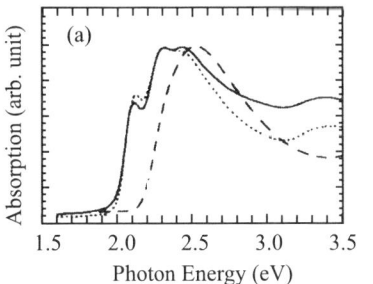
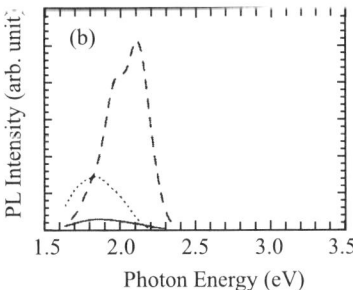

FIG. 5.4. (a) Absorption and (b) photoluminescence spectra from polymer films. The dashed line is for MEH-PPV, the dotted line is for CN-PPV and the solid line is for CN-PPV:MEH-PPV in 1:1 ratio [34].

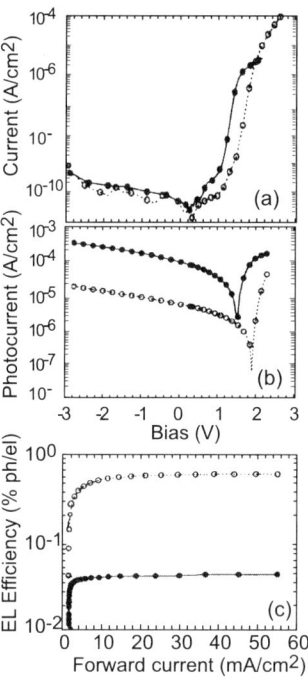

FIG. 5.5. (a) Dark current, (b) photocurrent under 430 nm 20 mW/cm^2 light illumination. (c) The electroluminescence efficiency. The solid circles are for the Ca/MEH-PPV:CN-PPV/ITO device and the open circles are for the Ca/MEH-PPV/ITO device [34].

circuit voltage was 1.6 V for the pure MEH-PPV cell and it decreased to 1.25 V for the composite cell. The dark and photocurrents of the cells fabricated by Halls et al. [122] showed similar behavior. The dark currents showed a rectification ratio of 10^3 at ± 3.5 V. Open circuit voltage was 0.6 V and short circuit current has a quantum efficiency of 6%. It increased to 15% for a reverse bias of 3.5 V and 40% at a reverse bias of 10 V. The values were even higher under forward bias. Fig. 5.5(c) shows the quenching of EL, a result consistent with that obtained by Halls et al. [122].

Yoshino et al. [35] fabricated two structures. In one structure D and A layers were OOPPV and C_{60} and the middle layer was OEP. In the second structure there was only one active layer consisting of the mixture of CN-PPV and PAT6. The electrodes were Al and ITO deposited on a glass substrate. Some structures were made on quartz substrates also. The absorption spectrum of the 60 nm OEP layer is shown in Fig. 5.6. The absorption spectra of the two layered [OOPPV(120 nm)/C_{60}(10 nm)] and the three layered structure [OOPPV(120 nm)/OEP(50 nm)/C_{60}(50 nm)] are also shown. The multilayers show strong absorption in the short wavelength region. The PL efficiency of CNPPV:PAT6 and CN:PPV:PDPA-TPSi blends were measured. In both cases the efficiency decreased considerably in the mixture. Gregg [123] found similar quenching of the PL in the PPEI layers if a thin TPD layer is inserted between the PPEI and the ITO. The TPD solution was spin coated onto the ITO film. The structures of the polymers used in Ref. [35] are shown in Fig. 5.7. Solar cells using the blend structure were made. Typical value of the

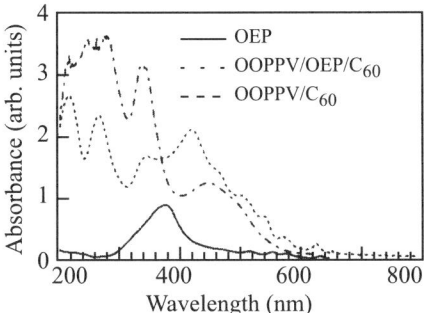

FIG. 5.6. Absorption spectra of the 60 nm OEP middle layer, the OOPPV/OEP/C_{60} composite layer, and the OOPPV/C_{60} layer [35].

open circuit voltage was 1.7 eV. The typical bandgap values are 2 eV and therefore the open circuit voltage is determined by the asymmetry of the contact electrodes [35].

PCBM (C_{61}-Butyric Ester) has been studied extensively. Recently a lot of work has been done on bulk heterojunction solar cells fabricated using PCBM and polymer acceptor–donor pairs [124,129, and references given therein]. Thin layer of LiF inserted between the cathode and the active layer has proved very useful in improving the performance of the organic LEDs. Brabec et al. [124] have studied the effect of inserting a thin (<15 Å) LiF layer between the Al electrode and the active layer of the BHSC. The active layer was formed using the MDMO-PPV/PCBM BH junction layer. The insertion of thin LiF layer increased the short circuit current and the Fill Factor significantly. To investigate whether the effect was due to the insulating nature of LiF, the authors fabricated and studied BHSCs with SiO_2 layer instead of the LiF layer. The improvement was not observed with the oxide layer. The authors concluded that the LiF layer lowers the work function of the electrode. They have discussed possible mechanisms of this reduction. A high conversion efficiency of 3.2% was obtained on inserting the LiF layer (see Table 5.1).

A very high efficiency, 3.5%, has been reported by Padinger et al. [129]. Padinger et al. [129] fabricated BHSCs using P3HT and PCBM as donor–acceptor pairs. The chemical structure of P3HT is shown in Fig. 5.8 and that of PCBM is shown later. They improved the performance of the solar cells by using a high temperature tempering cycle and simultaneously applying an external voltage. There was considerable improvement in the short circuit current, open circuit voltage and external quantum efficiency (see Table 5.1). The highest efficiency reported so far is 5% [130]. At these efficiencies the cells are cost effective and can be manufactured commercially. However long term stability continues to be a problem. Some work on accelerated testing has been done. At 85 °C the efficiency decreases by 20% in 600 h [130].

Kymakis et al. [131] have fabricated the BHSCs using carbon Single Wall Nanotubes (SWNTs) and P3OT polymer. TEM investigations show that SWNT powder contains 1.4 nm diameter tubes self-organized into bundle-like crystallites. Absorption spectra of the P3OT-SWNTs alloys were measured with different concentrations of the nanotubes. For low concentrations of the nanotubes there is not much change in the absorption spectrum. This indicates that there is no significant interaction between the two materials and

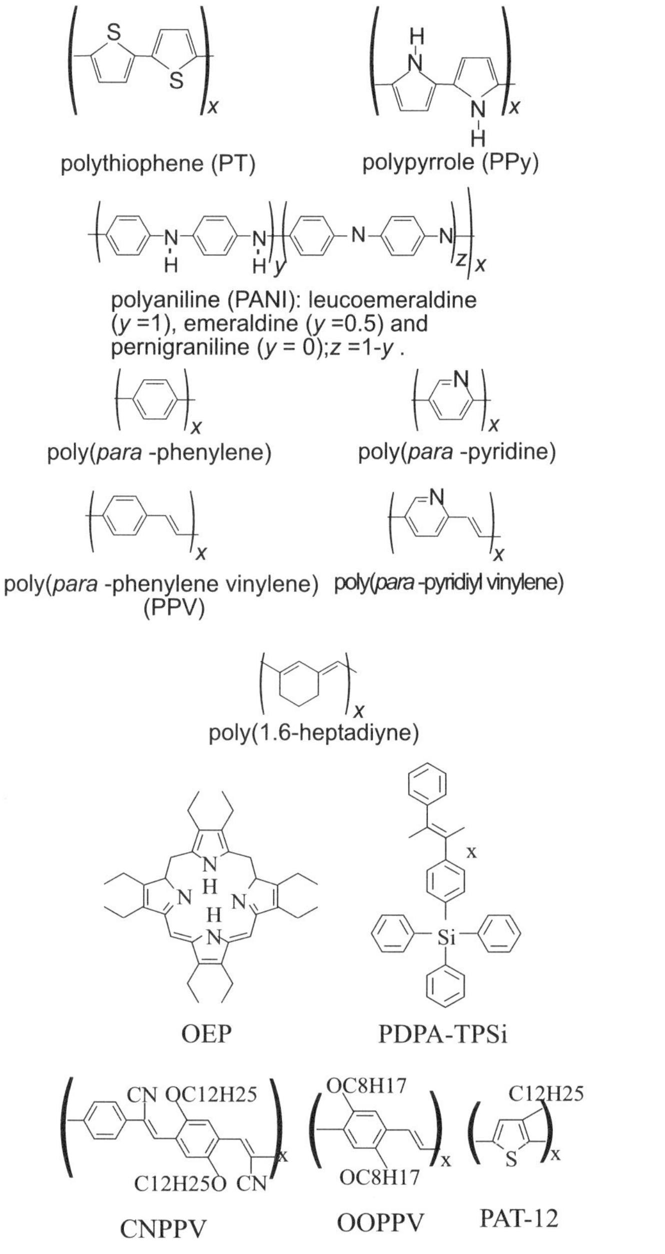

FIG. 5.7. Structure of some organics used for fabricating photovoltaic devices [35,129].

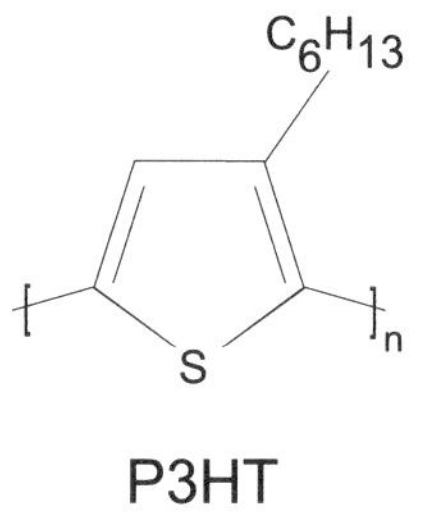

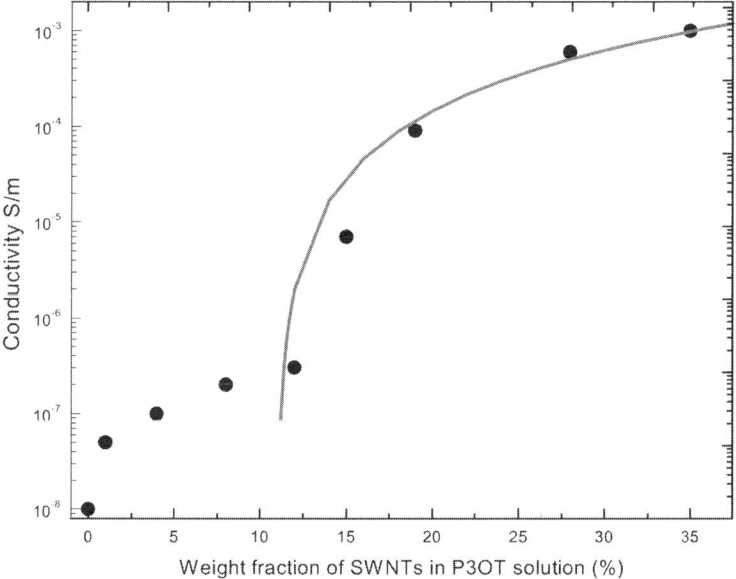

FIG. 5.8. Chemical structure of poly(3-hexyl thiophene) P3HT [129].

no charge transfer takes place in the ground state. From the absorption edge of the samples with low nanotube concentration a value of 2.4 eV for the band gap of P3OT was derived. There is an increase in absorbance at low and high energies due to the metallic nature of the nanotubes. The decrease in absorbance at the peak in the high nanotube concentration samples is observed due to the reduced volume of the P3OT.

The conductivity of the alloys as a function of nanotube concentration is shown in Fig. 5.9. P3OT is an insulator with a conductivity 10^{-8} S/m. The conductivity increases monotonically with nanotube concentration. However the rate of increase is relatively small at low concentrations. It increases dramatically above 12% concentration. Above this concentration the nanotubes are close enough so that percolation becomes possible. The fit of the percolation theory is shown by the solid line.

FIG. 5.9. Electrical conductivity variation versus weight fraction of SWNTs in the blend (filled circles). The percolation threshold is found to be 11%. The solid line is the fit of the percolation model [132].

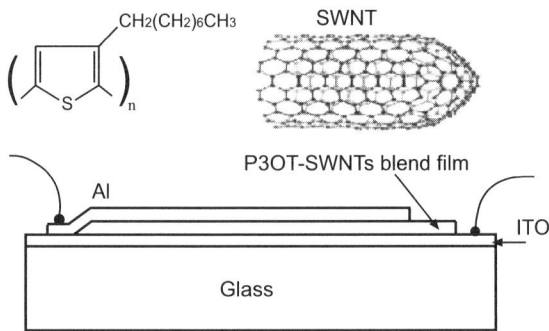

FIG. 5.10. Left: Chemical structures of P3OT, SWNTs and device architecture of the photovoltaic cell [131].

A schematic diagram of the BHSC and structure of P3OT and SWNT are shown in Fig. 5.10. The concentration of SWNT was 1%. The films were prepared by the drop and spin coating on ITO coated quartz substrate. The other electrode was Al deposited by thermal evaporation. The characteristics of the devices were measured in the dark and under illumination by the light equivalent of 1.5 AM solar spectrum. The current is considerably higher in the forward bias than in the reverse bias, which suggests that the devices are good quality rectifying diodes. The illuminated characteristics show that the short circuit current increases considerably in the alloy devices. The short circuit current density was 0.12 mA/cm^2 and the open circuit voltage was 0.75 V. The corresponding quantities in the pristine P3OT device were 0.7 µA/cm^2 and 0.35 V. Here the emphasis is on the improvement in the BHSC and not on the actual performance. The actual performance will certainly improve with improvement in the technology.

5.3. Source of V_{oc} in BHSCs

5.3.1. EFFECT OF ACCEPTOR STRENGTH [133]

Until recently it was not clear whether the V_{oc} of the polymer:fullerene BHSCs is determined by the work function difference of the electrodes or by the internal junctions of the two constituent materials. Brabec et al. [133] synthesized a series of highly soluble fullerene derivatives with varying acceptor strengths (i.e., first reduction potentials). These fullerene derivatives, methanofullerene PCBM, a new azafulleroid and a ketolactam quasifullerene are shown in Fig. 5.11. Redox behavior of all the four acceptors was studied using cyclic voltammetery. Solubilizing groups have comparable sizes. Effects due to different donor–acceptor distances, and/or different morphologies, are minimized. The electron acceptor strength of the compounds is different. Ketolactam 6 (± 0.53 V) appeared to be a substantially better electron acceptor than C_{60} (± 0.60 V), whereas azafulleroid 5 (± 0.67 V) is close to C_{60}, and PCBM (± 0.69 V) showed a clearly diminished electron affinity. A difference of 160 mV is observed between the strongest and the weakest acceptor.

Thin film BHSCs were fabricated [133] with MDMO-PPV/fullerene-derivatives. The active layers were about 100 nm thick. The authors also fabricated cells with fullerene

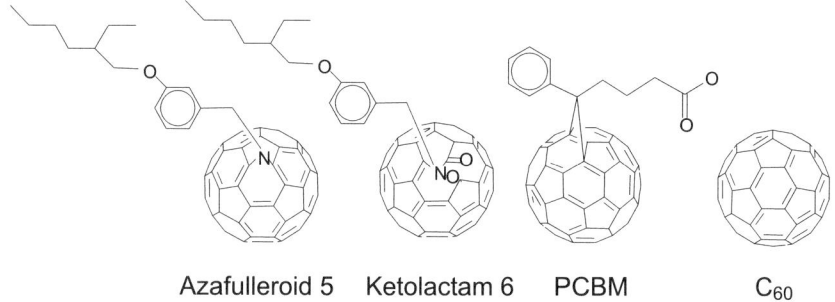

FIG. 5.11. Chemical structures of the investigated compounds [133].

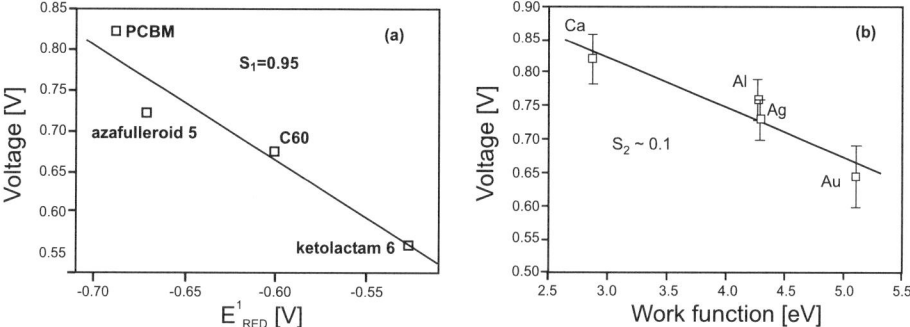

FIG. 5.12. (a) V_{OC} versus acceptor strength and (b) V_{OC} versus negative electrode work function. The S_1 and S_2 are the slopes of the linear fits to the data [133].

(C_{60}) for comparison. For the three acceptors presented, a relatively narrow distribution of the open circuit voltage was observed, indicating excellent reproducibility: Four metals with different work functions W [Ca ($W = 2.87$ eV), Al ($W = 4.28$ eV), Ag ($W = 4.26$ eV), and Au ($W = 5.1$ eV)] were used as the negative electrodes. A total variation of less than 200 mV of the V_{oc} was observed for a variation of the negative electrode work function by more than 2.2 eV. This shows that the open circuit voltage depends weakly on the electrode work function. The measured V_{oc} in the cells fabricated with the three different derivatives are shown in Fig. 5.12. This figure shows that the V_{oc} is strongly dependent on the derivative used. The authors [133] concluded that the V_{oc} depends mainly on the internal bulk heterojunction.

5.3.2. MORE RECENT WORK [134,135]

Mihailetchi [134] investigated the open circuit voltage of the bulk heterojunction organic solar cells based on methanol-fullerene [6,6]-phenyl C61-butyric acid methyl ester (PCBM) as electron acceptor and poly[2-methoxy-5(3',7'-dimethyloctyloxy)-p-phenylene vinylene] (OC_1C_{10}-PPV) as an electron donor. It is known that a single layer device follows the MIM model [166] and the open circuit voltage V_{oc} is equal to the difference in the work functions of the metal electrodes [134]. If charges accumulate in the

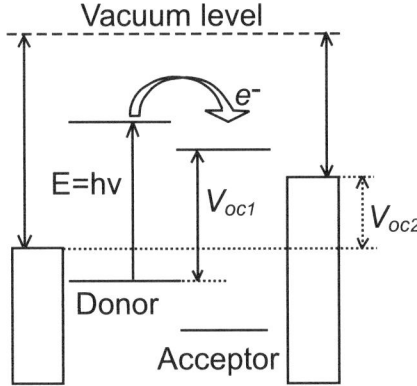

FIG. 5.13. Schematic variation of V_{oc} with acceptor strength (solid double headed arrow, V_{oc1}) or/and electrode work function (dotted arrow, V_{oc2}), in a donor/acceptor BHJ solar cell. The electron transfer, occurring at the donor/acceptor interface after light excitation, is indicated by the bent arrow [134].

organic layer and the metal electrode, they diffuse away. A drift current is created to balance this diffusion. The resulting field produces a voltage drop, which is subtracted from the work function difference to obtain the observable open circuit voltage. The positive electrode was a ITO coated glass and four different materials, LiF:Al, Ag, Au, and Pd, were used for cathodes. The V_{oc} was different in each case, it being minimum for Pd and maximum for LiF:Al. This result is in direct contradiction of the result of Brabec [133] discussed earlier.

If the electrodes match the lowest unoccupied molecular orbital (LUMO) of the acceptor and the highest occupied (HOMO) level of the donor, respectively, the contacts can be regarded as ohmic. The maximum V_{oc} for this case is schematically indicated by V_{oc1} in Fig. 5.13 and is thus controlled by the bulk active layer material properties. Non-ohmic contacts, as shown in Fig. 5.13, a V_{oc} with magnitude V_{oc2} should be observed, according to the MIM model. However if the Fermi level of the contact metal is pinned at the LUMO and of the anode with the HOMO, the observed V_{oc} by the properties of the acceptor and the donor and will become insensitive to the work function difference of the electrodes.

Ramsdale et al. [135] have investigated the open circuit voltage of a bilayer solar cell. The electron-accepting polymer used was poly 9,9'-dioctyl-fluoreneco-benzothiadiazole) (F8BT) and the hole-accepting polymer was poly-(9,9',dioctylfluorene-co-bis-N,N'-(4-butylphenyl)-bis-N,N'-phenyl-1,4-phenylenediamine) (PFB) (see Fig. 5.14). The anode was ITO coated glass in all cases. Five different cathodes shown in Table 5.2 were used. Illuminated characteristics were measured with the light of 459 nm wavelength and intensity of incident light was 0.7 mW/cm^2. The work-function difference of the two electrodes is compared with the measured V_{oc} in Table 5.2.

Table 5.2 compares the measured open-circuit voltages with the difference in the work functions ($\Delta\phi$) of the electrodes. The table shows that with the exception of calcium, the open-circuit voltage increases linearly with the work function difference. However there is an additional constant 1 V contribution to the open-circuit voltage that cannot be accounted for by the difference in work functions. Calcium work function is smaller

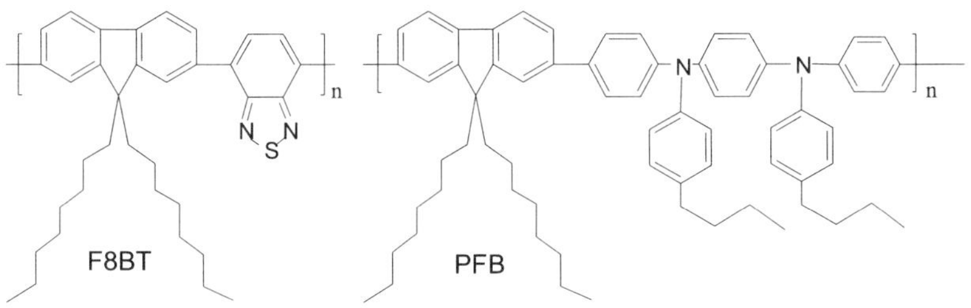

FIG. 5.14. Chemical structures of F8BT and PFB [135].

TABLE 5.2
WORK FUNCTIONS ϕ OF THE ANODES AND CATHODES USED AND THE COMPARISON BETWEEN WORK
FUNCTION DIFFERENCE ($\Delta\phi$) AND OPEN CIRCUIT VOLTAGE (V_{OC}) [135]

Anode	Cathode	$\phi_{cathode}$ (eV)	$\Delta\phi$ (eV)	V_{oc} (V)	$V_{oc} - \Delta\phi$ (V)
ITO	gold	5.1	−0.3	0.7	1.0
ITO	copper	4.65	0.15	1.15	1.0
ITO	chromium	4.5	0.3	1.35	1.05
ITO	aluminum	4.3	0.5	1.5	1.0
ITO	calcium	2.9	1.9	1.5	−0.4

The work function of ITO was taken to be 4.8 eV.

than the electron affinity of F8BT, leading to charge transfer at the interface, and pinning of the electrode work function close to the energy of the LUMO of the polymer. The excitons dissociate and the carriers are created near the interface. Due to concentration gradient they diffuse away from the interface. Since net current must be zero at every point in the open circuit configuration, an electric field is created which gives rise to a drift current to balance the diffusion current. The electric field produces a voltage within the active layer, which accounts for the additional voltage given in Table 5.2. However a strong dependence of the open circuit voltage on the work function of the electrodes remains.

5.4. Optimum PCBM Concentration

Van Duren et al. [136] made extensive investigations of the morphology of PCBM and MDMO-PPV using TEM and AFM techniques. To study the morphology at different depths, a deuterated derivative of PCBM (d5-PCBM) was used. The structure of the chemicals used are shown in Fig. 5.15. AFM, TEM and dynamic SIMS techniques were used. Deuterium labeling of PCBM (d5-PCBM) combined with dynamic SIMS gives morphology at different depths [136]. AFM results are shown in Fig. 5.16. The thickness of the films was ~100 nm. AFM heights and corresponding phase images are shown for four compositions. Several other compositions were studied but the results for the other compositions are not shown in the figure. The surfaces are extremely smooth for PCBM concentration smaller than 50%. The heights (maximum peak to valley roughness) were

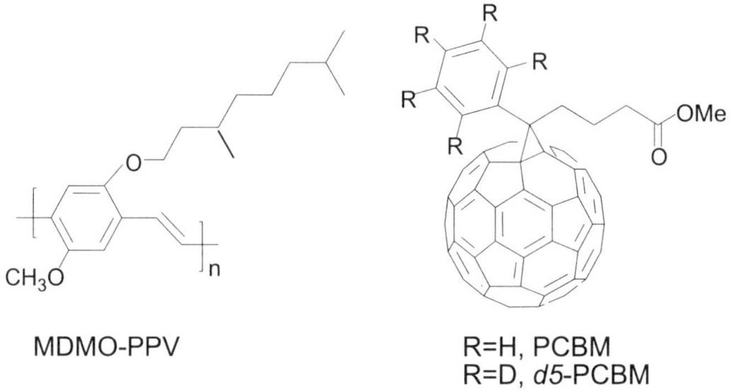

FIG. 5.15. Molecular structure of MDMO-PPV, PCBM and d5-PCBM [136].

3 nm and root mean square (RMS) values were 0.4 nm for 2 μm × 2 μm area films. As the PCBM concentration increases from 67 to 90%, surface roughness increases rapidly. The values of the height increase from 3 to 22 nm and the RMS values from 0.4 to 3.3 nm. Simultaneously a reproducible phase contrast appears. Separate domains of one phase in a matrix of another phase can be recognized. The domain size increased from 40–65 nm for 67% PCBM to 110–300 nm for 90% PCBM. The domain size depended on the thickness of the films. For 80% PCBM, the domain size was 60–80 nm for 65 nm films and it was 100–120 nm for 270 nm films. The main result of AFM studies that below 50% PCBM concentration, the PCBM is distributed molecularly or as very tiny particles in the MDMO-PPV matrix and above 67% phase separation of PCBM takes place on nanoscale is very fully confirmed in depth by dynamic SIMS measurements.

TEM results are shown in Fig. 5.17. Though the TEM measurements were made for several concentrations in the range 0–100% PCBM, TEM pictures only at four compositions are shown in the figure. AFM shows the structure at the surface whereas the TEM reveals the structure in the bulk of the sample. TEM results are in complete agreement with the AFM results. Phase separation occurs above 67% PCBM and increases as the PCBM concentration increases further. Below 50% PCBM the PCBM and MDMO-PPV form homogeneous mixture.

Optical measurements showed that a small concentration of 2% is effective in quenching the PL of MDMO but a high concentration of more than 67% is required to increase the photoconductivity or the short circuit current of a BHJSC. The solar cells efficiency measurements show that the optimum PCBM concentration is 80%. See next section.

5.4.1. SUPERPOSITION PRINCIPLE

The superposition principle has been discussed by several authors [137–139] and has been reviewed by Jain [140]. The differential equations describing the charge densities and currents in the solar cell are linear and can be superposed. The superposition leads to the result,

$$I(V) = I_d(V) - I_{sc}. \tag{5.1}$$

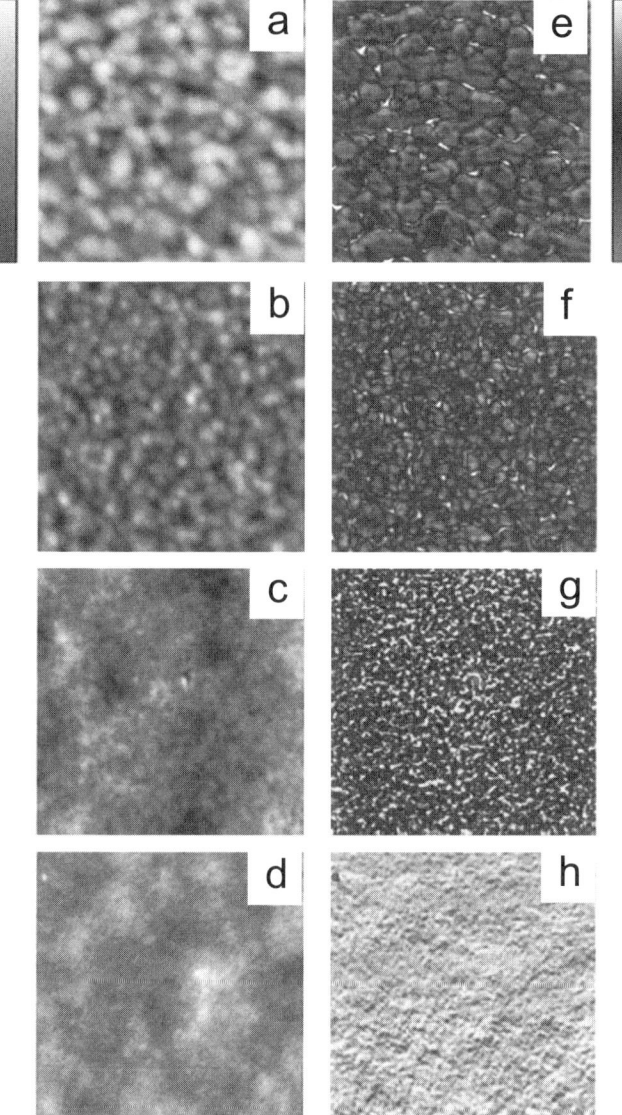

FIG. 5.16. AFM height (a to d) and simultaneously taken phase (e to h) images of the MDMO-PPV/PCBM composite films of 90 (a, e), 80 (b, f), 67 (c, g), and 50 wt% PCBM (d, h). The height bar (maximum peak-to-valley) represents 20 nm (a), 10 nm (b), 3 nm (c), and 3 nm (d). The size of the images is 2 μm × 2 μm [136].

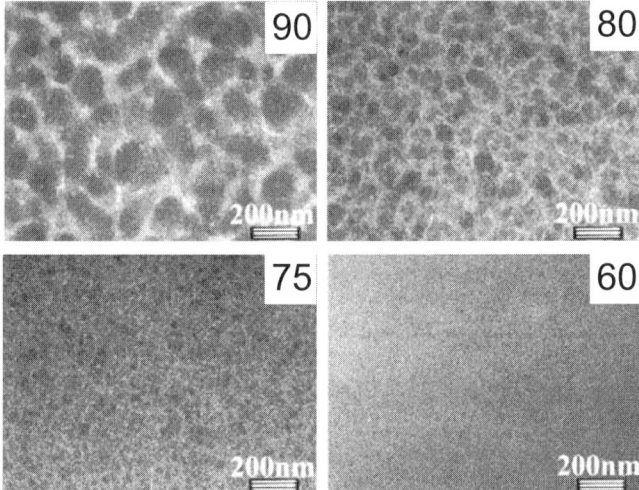

FIG. 5.17. TEM images of MDMO-PPV/PCBM blends with different weight percentages of PCBM as indicated on the upper right corner [136].

In the open circuit configuration $I(V) = 0$ and $I_d(V) = I_d(V_{oc}) = I_0 \exp(qV_{oc}/kT)$. We obtain,

$$V_{oc} = \frac{kT}{q} \ln \frac{I_{sc}}{I_0}. \qquad (5.2)$$

Here I_0 is the dark saturation current. Eqs. (5.1) and (5.2) are valid under the following assumptions [139].

1. Series and shunt resistances are negligible.
2. There is no contribution from the space charge layer to the dark or the short circuit current.
3. There are no trapping centers.
4. High injection conditions do not prevail.
5. Material parameters (e.g. mobility μ) and electric field distribution do not change under illumination.

Experiments show that in high quality Si solar cells the superposition principle is valid to a good approximation. In CdS/Cu$_2$S, amorphous Si [139] and in polymer solar cells some of these approximations are grossly violated. As an example consider the effect of series resistance. In the presence of the series resistance R_s the dark current is given by,

$$J(V) = J_s \left[\exp\left(\frac{q(V - JAR_s)}{kT} \right) - 1 \right] - J_{sc}. \qquad (5.3)$$

The value of J in the exponent is different in the dark and in the illuminated situations. This invalidates the superposition principle. It can be easily demonstrated that shunt resistance will also invalidate the superposition principle. Under high intensity of illumination material parameters can change and also due to the different mobilities of electrons and

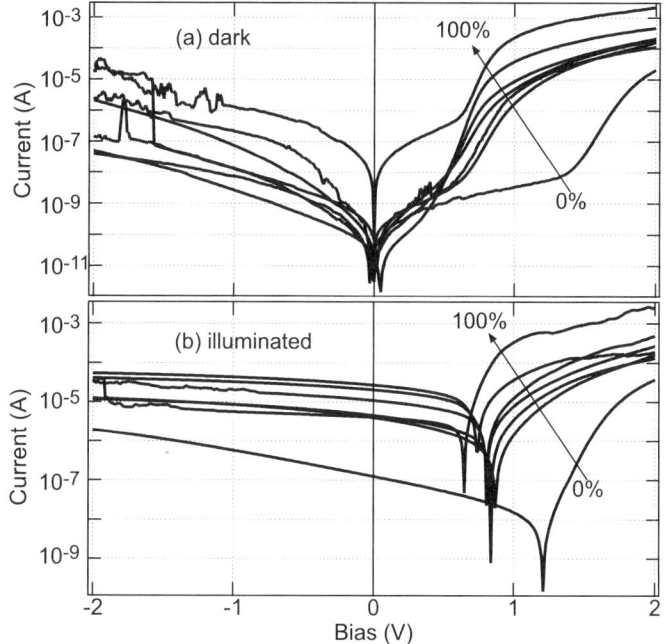

FIG. 5.18. Variation of dark and illuminated current with voltage for ITO/PEDOT/MEH-PPV:PCBM/Al blends for different concentrations of PCBM. The contacts were ITO and Al [43].

holes, an electric field is created. To make corrections for these effects in the superposition principle for polymer cells is extremely difficult.

The dark and illuminated characteristics of the PCBM:PPV solar cells for different concentrations of PCBM are shown in Fig. 5.18. It can be seen that the dark characteristics are not described by Shockley type equation. Forward illuminated $I(V)$ curves are strongly dependent on the applied voltage. Under reverse bias the illuminated current is not so sensitive to the voltage.

The short circuit current I_{sc} and the open circuit voltage V_{oc} are shown in Fig. 5.19. It is seen that the short circuit current is strong function of the composition. The open circuit voltage is also dependent on PCBM concentration. It is difficult to conclude whether or not open circuit voltage is entirely dependent on bulk heterojunctions. The open circuit voltage depends on short circuit current and also on dark saturation current. The dependence of dark saturation current on PCBM concentration is not known. It is clear from the figure that the optimum PCBM concentration is 80% where the short circuit current is maximum.

We have used a simplified approach to check whether the superposition principle is approximately valid. We apply Eq. (5.1) directly to the experimental results obtained with the solar cells. We choose three values of the voltage, $V = V_{oc}$, $V = V_p$ (V_p is the voltage at the maximum power point) and $V = 0.25 V_{oc}$. We read I_d and I_{sc} for each voltage from Fig. 5.18. We do this for each composition. We then plot $\delta I = I_d - I_{sc}$ versus composition. We also plot $I(V)$ read from Fig. 5.18. The difference of δI from $I(V)$ is a measure of the deviation from the superposition principle. For open circuit

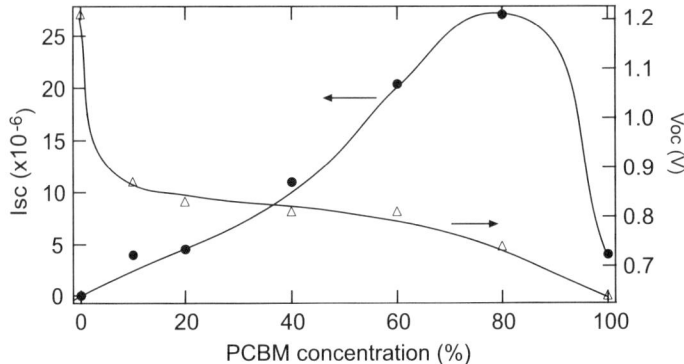

FIG. 5.19. The plots of V_{oc} and I_{sc} are shown as a function of composition of the sample [43].

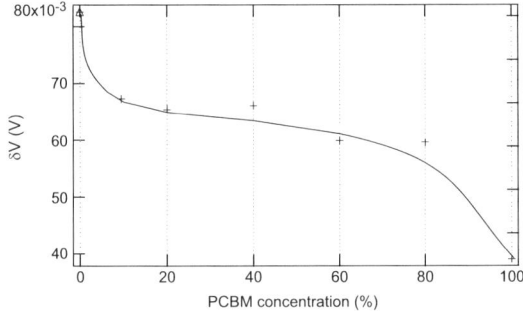

FIG. 5.20. Plot of δV as a function of PCBM concentration [43].

configuration this result can also be expressed as [43]:

$$\delta V = \frac{kT}{q} \ln \frac{I_l}{I_d(V = V_{oc})}. \tag{5.4}$$

If superposition principle is valid, δV is zero. Therefore the value of δV is a measure of the deviation from the superposition principle. The plot of Eq. (5.4) is shown in Fig. 5.20. The values of δV are small and the discrepancy is not very large.

5.5. Modeling the Output Characteristics

5.5.1. THE OUTPUT CURRENTS

The dark and illuminated $J-V$ curves of the solar cells for different PCBM concentrations are plotted in Fig. 5.21. Measurements for 30% and 60% PCBM were also made but they are not shown as they fit with the general trends shown by the curves for other concentrations. The features of the $J-V$ curves in Fig. 5.21 are mentioned below. (1) For larger voltages the dark currents are space charge limited currents. The shape of the dark currents, particularly the points of inflexion, are characteristics of the space charge limited

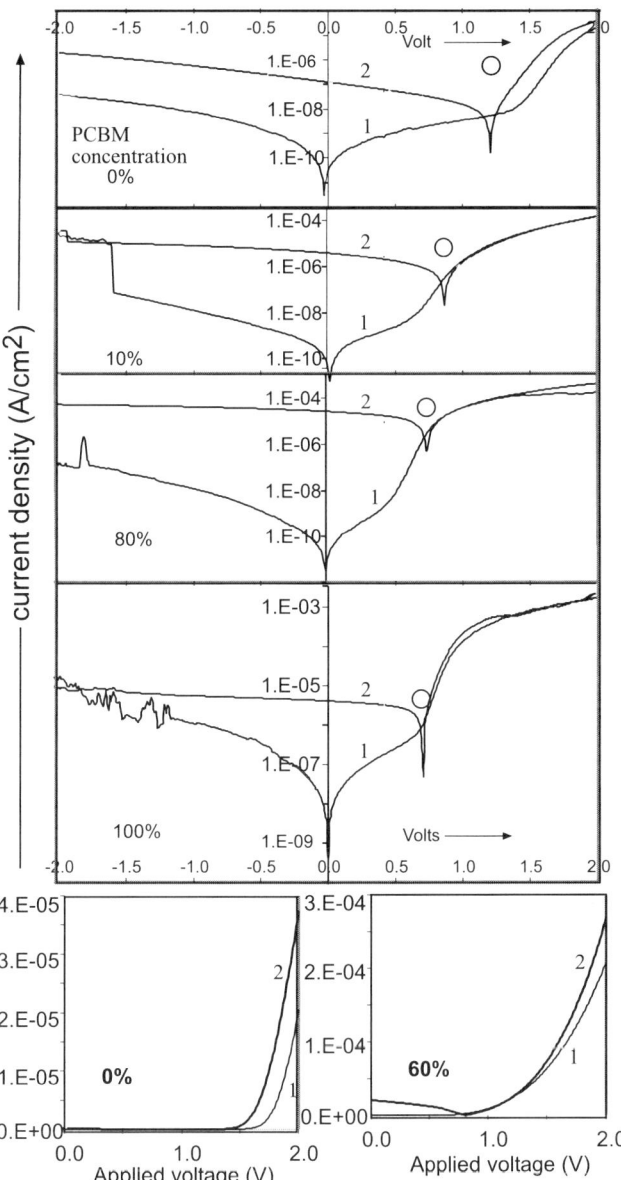

FIG. 5.21. Top figure: The experimental plots of dark and illuminated J–V characteristics of a typical MDMO-PPV/PCBM solar cell for different concentrations of PCBM. The currents are plotted on the log scale. Curves 1 are for dark currents and curves 2 are for illuminated currents. The open circles show the dark currents needed to make the output current zero at the open circuit voltage. Bottom figure: The current densities are plotted on the linear scale for two PCBM concentrations. Curves 1 are for dark currents and curves 2 are for illuminated currents.

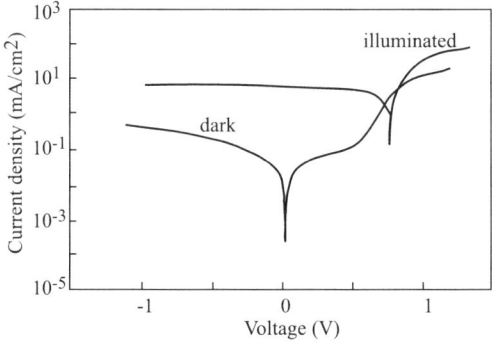

FIG. 5.22. Experimental plots of $I-V$ characteristics of a typical MDMO-PPV/PCBM bulk heterojunction solar cell with a LiF/Au electrode [124].

currents [43,41,46]. In the small voltage range the currents are either the leakage currents or the ohmic currents due to background doping. (2) The shapes of the illuminated curves do not resemble to that of the dark currents. For small concentrations of PCBM the reverse illuminated current increases with the applied negative bias. For high PCBM concentrations the reverse illuminated characteristic becomes practically flat. The increase is most probably due to field enhanced dissociation of the excitons since it does not occur at larger PCBM concentrations. (3) An important feature of the curves in Fig. 5.21 is that at higher forward voltages the dark and illuminated currents do not differ by appreciable amounts. In fact in many samples the dark current intersects the illuminated current and becomes smaller than the illuminated current (see the linear plots for 0 and 60% PCBM at the bottom of the figure). This suggests that on applying forward bias the nature and magnitude of the dark current change significantly under illumination. (4) At open circuit voltage the dark current must always be equal to the short circuit current. We have shown the value of the dark current expected from this criterion by the large open circles. It is seen that under illumination the measured dark current is smaller than its expected value by a factor between 5 and 10.

Brabec and coworkers have made extensive measurements of the dark and illuminated currents of MDMO-PPV/PCBM BHSCs [124,141]. The experimental results for the BHSCs from their paper [124] are shown in Fig. 5.22. There are points of inflexion in the dark current of the organic solar cell confirming that the dark current is the space charge limited currents. These features, particularly the points of inflexion, are absent in the illuminated characteristics. At larger voltages the forward dark current intersects the illuminated current and the illuminated current becomes larger than the dark current.

In 1995 Yu et al. reported $J-V$ characteristics of a BHSC (layer structure Ca/MEH-PPV:PCBM/ITO) device in their classical papers [119]. Dark forward current had the usual shape with a point of inflexion characteristic of the space charge limited currents. However the dark forward current was smaller than the illuminated curve at all voltages. The result is not acceptable, for if this were true the observed illuminated current should have flown in the opposite direction, i.e. it should have been negative like the short circuit current. In their paper in Science [119] Yu et al. observed similar discrepancy in the single layer organic solar cells also.

Yoo et al. [142] fabricated organic solar cells based on bilayer heterojunctions of pentacene and C_{60}. The external quantum efficiency of the device was high because of large diffusion length of the exciton in pentacene and because of the efficient dissociation of the excitons at the heterojunctions. The dark and illuminated characteristics of the cells were plotted. Here also the dark and illuminated curves cross. Since this paper is not on BHSCs, it shows that the peculiar features of the I–V plots we have discussed are the general features of the organic solar cells and are not limited to the bulk heterojunction structure of the BHSCs. Pfeiffer et al. [143] have plotted in their Fig. 9 the observed characteristic of heterojunction between highly doped zinc-pathalocyanine (ZnPc) (30 mol% tetrafluoro-tetracyano-quinodimethane (F4-TCNQ)) and N,N'-dimethyle-3,4;9,10-tetracarboxilic-diimide (MePTCDI) in the dark and under white illumination, measured in-situ in high vacuum at room temperature. Crossing of the dark and illuminated currents at a rather small voltage is clearly seen. Several other workers [144,145] have obtained similar results. In fact we did not find any paper where the dark current continues to be sufficiently larger than the illuminated current at higher voltages.

To summarize this section, we can say that the measurements made by us and by many other groups appear to be inconsistent with general physical concepts. Though this behavior has been observed in solar cells fabricated with different organic materials and with different structures, the polymer scientists have not discussed reasons for this behavior.

5.5.2. THE MODEL

Earlier we have modeled the dark forward characteristics of the BHSCs and other organic solar cells using the theory based on space charge limited currents (SCLC) [1,3]. We were not successful in modeling the illuminated characteristics of the cells using the SCLC theory. We present a model, which allows the interpretation of the output characteristics of the illuminated organic solar cell. In our earlier work [40] we have shown theoretically and experimentally that if there are free carriers due to back ground doping and if the injected carrier density is smaller than the doping induced carrier density, the space charge limited currents do not exist. We [30] have shown that a similar situation arises in an illuminated organic solar cell. At one sun illumination the light induced carriers are larger in concentration than the injected carrier density over most of the region. Due to large free carrier density all the traps are filled and do not play a significant role in the transport of carriers. *Therefore the forward current due to a voltage existing on the junction (under illumination) is quite different from the measured dark current. It is not the measured dark current but the current calculated using Shockley equation* $J = C\{\exp(q[V - AJR_s]/nkT) - 1\}$ *where A is the area which is more relevant.*

A dark forward current I_d flows in the cell if a voltage is applied to the junction. To avoid confusion and for the sake of clarity we will call this current a junction current I_j measured in the dark, i.e. $I_d = I_j(\text{dark})$. This 'dark current' continues to flow even if the cell is illuminated. The only requirement for the current to flow is that there should be a voltage at the junction. We designate this current measured under illumination as $I_j(\text{illumination})$. We need to introduce this notation because unlike in Si solar cells, in organic solar cells $I_d = I_j(\text{dark})$ is not equal to $I_j(\text{illumination})$. The junction current density of a Si solar cell measured in the dark is given by the diode equation [45],

$$J_j(\text{dark}) = J_0\left[\exp(q[V - J_j A R_s]/nkT) - 1\right] + (V - J_j A R_s)/R_p. \tag{5.5}$$

Here J_0 is the dark saturation current density, R_s is the series resistance, A is the area and R_p is the shunt resistance. When a voltage V exists on the junction, the output current of the illuminated solar cell has two components, (1) the junction current density J_j(illumination) due to injection of carriers and (2) the short circuit current density J_{sc} due to light generated carriers. Therefore the output current density J_{lv} of the illuminated solar cell is given by,

$$J_{lv} = J_j(\text{illumination}) - J_{sc}, \tag{5.6}$$

where

$$J_j(\text{illumination}) = J_0\left[\exp(q[V - J_j A R_s]/nkT)\right].$$

The negative sign on the right-hand side of (5.6) indicates that the light generated current flows in a direction opposite to that of the junction current. For a good Si solar cell $J_j(\text{illumination}) = J_j(\text{dark})$. This gives the well-known superposition principle implicit in Eq. (5.6). For an organic solar cell the junction current density $J_j(\text{dark})$ is space charge limited and is given by,

$$J_j(\text{dark}) = \text{SCLC}, \tag{5.7}$$

where SCLC indicates a space charge limited current. The SCLC varies as V^2 in a trap-free insulator and in a more complex manner if traps are important.

In the case of good quality Si solar cell the superposition principle holds. Eq. (5.6) remains valid with the value $J_j(\text{illumination}) = J_j(\text{dark})$ and the value of J_{sc} measured at 0 voltage remains valid at other voltages. It is not likely to be the case for the organic solar cells. The $J_j(\text{dark})$ will change on illumination to $J_j(\text{illumination})$ and J_{sc} may change on the application of the voltage. The changes must be large to make $J_j > J_{lv}$ and to make the separation between J_{lv} and $J_j(\text{illumination}) = J_{sc}$. Earlier work on polymer solar cells [124] and on amorphous solar cells [146,147] suggests that the modification in the short circuit current by the light cannot be very large. In amorphous solar cells the nature and magnitude of the junction current can change drastically on illumination [146,147].

Our model is based on these observations; it is simple and capable of interpreting all known experiments. The model assumes that the space charge in the illuminated organic solar cell is cancelled by the relaxation of the large density of light induced free carriers; much in the same way as the majority carriers cancel the space charge due to injected minority carriers in a Si p–n junction diode. The junction current is now determined by diffusion and recombination and can be described by the diode equation, Eq. (5.5). The space charge limited dark current shown by curves 1 in Fig. 5.21 changes to exponential form, Eq. (5.5), under illumination. The 'dark junction current' in organic solar cells measured in the dark, Eq. (5.7), has no practical significance because the solar cells in use are always under illumination. The I–V characteristics of the illuminated organic solar cell are calculated using Eqs. (5.5) and (5.6). The theory of illuminated organic solar cell now becomes similar to the theory of a Si solar cell.

The fit of the model to the experimental data of Fig. 5.21 for 80% PCBM, is shown in Fig. 5.23. The values of the parameters used are given below the figure. We found

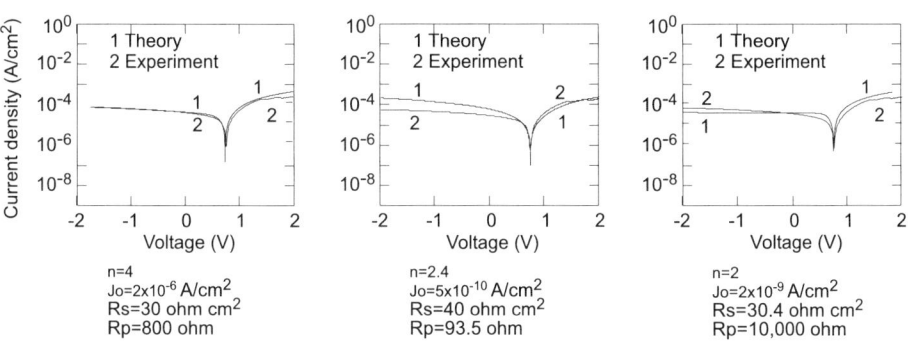

FIG. 5.23. Fitting of the model with experimental results for three different sets of parameters.

that the best fit requires $n = 4$. Though values of n greater than 2 have been reported in the literature [148], a value $n = 4$ appears rather large, particularly since most traps are likely to be filled under illumination. We found that acceptable fits can be obtained for smaller values of n as shown in Figs. 5.21(b) and 5.21(c). The good fit with the model can be obtained for different sets of parameters. Similar plots were obtained with other concentrations of PCBM. We have compared our model with experimental data obtained with several organic solar cells fabricated by other groups and found good agreement between the model and the experiments. Again we could fit each data with more than one set of parameters. This weakness of the model can be removed only when more reliable values of the material parameters become available.

In conclusion we have shown that the junction current of an organic solar cell measured in the dark cannot be used to interpret data obtained with illuminated solar cells. The current changes from SCLC to diffusion and recombination current when the cell is illuminated. This has allowed us to provide a physical basis for the use of a model, which uses the diode equation for the illuminated solar cells. The model is very successful in interpreting the output characteristics of the illuminated organic solar cells by Brabec and confirmed by us. The true junction 'dark current' (diffusion and recombination current) determined by us remains more than the output current of the illuminated solar cell. Now the two currents do not intersect. This also explains why points of inflexion observed in the current measured in the dark disappear in the output current of the illuminated cell.

5.6. Comparison with Other Solar Cells

5.6.1. Amorphous Si Solar Cells

Han et al. [147] were interested in determining the built in potential in a-Si:H p-i-n solar cells. They measured extensively the I–V characteristics of dark and illuminated solar cells. The dark characteristics are shown in Fig. 5.24(a). The symbols are the experimental data and the solid lines are the fits of SCLC theory. The theory agrees with the experimental data well except for small voltages. As discussed earlier the discrepancy at small voltages is either due to background doping or due to leakage currents. Since

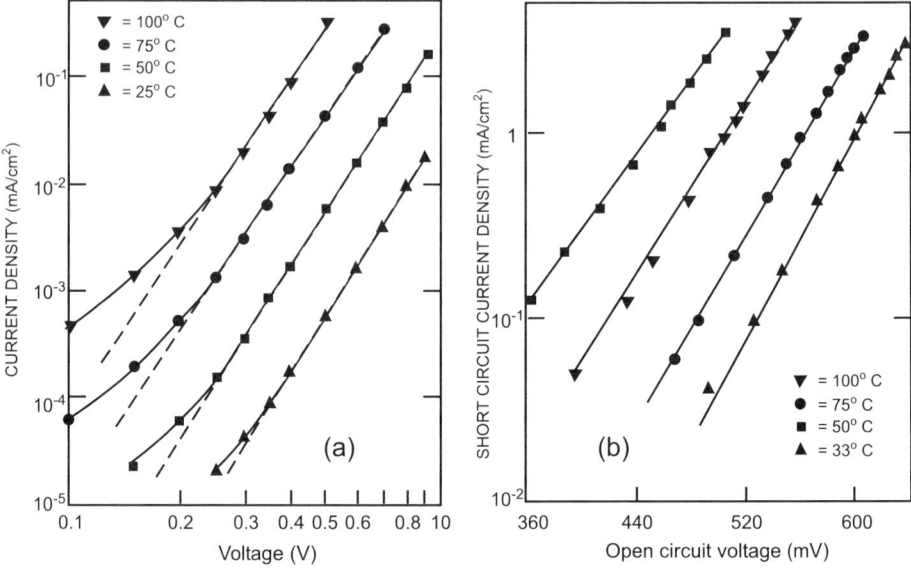

FIG. 5.24. (a) Fit of SCLC theory to the experimental data. The deviation at small voltages is due to back ground doping [147]. (b) Agreement of experimental data with Eq. (5.8) [147].

SCLC is bulk control in these materials, it can not yield any information about the built-in voltage at the contact or at the junction. Han et al. [147] showed that the illuminated characteristics closely obey Eq. (5.8) as shown in Fig. 5.24(b).

$$V_{\text{oc}} = \frac{kT}{q}\left[\ln\left(\frac{J_{\text{sc}}}{J_0}\right)\right]. \tag{5.8}$$

Eq. (5.8) is derived from Eq. (5.6) by using $V = V_{\text{oc}}$ and $J_{\text{lv}} = 0$. Eq. (5.6) is based on the principle of superposition and its derivation uses Eq. (5.5). Validity of Eq. (5.8) implies that the junction current (the so called dark current) under illumination is given by Eq. (5.5), a result used above by us for the organic solar cells.

Hegedus et al. [146] also measured the I–V curves of amorphous Si p-i-n solar cells at different temperatures. The curves in the dark and also under illumination are shown in Fig. 5.25. The I–V curves under illumination obey Eq. (5.8) in this case also. The built in voltage calculated from the illuminated curves was satisfactory. Hegedus et al. [146] took small portion of the dark I–V curves which could be represented by an equation of the type (5.5). They then attempted to derive the built in voltage using this dark current as Shockley currents. They obtained very strange results. Hegedus et al. [146] wrote: "Our results also clearly demonstrate that it is completely inappropriate to analyze $J(V)$ data measured in the dark on a-Si p-i-n devices and then attempt to correlate the results with photovoltaic performance.... Measurements in the dark are not applicable to understanding illuminated photovoltaic performance." It can be easily concluded from this work [146] that on illumination the SCLC changes to Shockley's diffusion current.

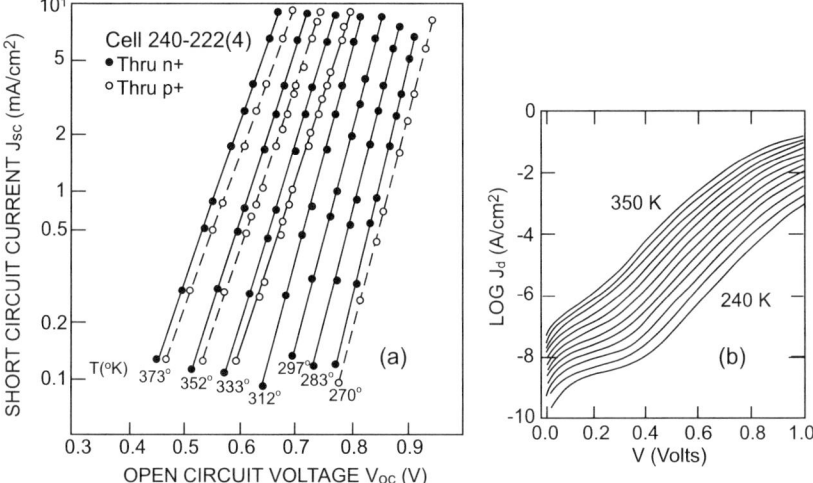

FIG. 5.25. Plots of (a) $J_{sc}(V_{oc})$ and (b) dark current $J_d(V)$ at different temperatures. In good quality single crystal Si solar cells J_{sc} and J_d are equal if $V_{oc} = V$. In amorphous Si solar cells $J_c \gg J_d$ proving that dark current increases on illumination [146].

5.6.2. POLYCRYSTALLINE SI SOLAR CELLS

The grain boundaries in poly-Si have a very large effect on most physical properties. Here we will investigate the effect of the light on the dark current. There are two major effects:

(1) In a certain range of doping, the grain boundary increases the resistance by several orders of magnitude. The dark current now becomes space charge limited. Under illumination the space charge is eliminated and the current changes from being space charge limited to Shockley type diffusion current.

(2) The grain boundaries reduce the life time and diffusion length of the minority carriers. The reduction is very sensitive to the illumination. Under illumination the barrier height at the grain boundaries decreases and the diffusion length increases. Since the diffusion length occurs in the denominator of the expressions for the dark current, the dark current decreases significantly. This increases the open circuit voltage and the fill factor of the solar cell. In addition to the normal increase of the efficiency in concentrated light, this is an additional effect in poly-Si. Our calculations show that this effect is quite significant.

The increase in resistance can be calculated. The grain boundary barrier height in the dark is given by [149],

$$\phi_{g0} = \left(\frac{qN_{ts}^2}{8\epsilon N}\right)\left[\frac{1}{1 + \frac{n_i}{N}\exp(q\phi_{g0} + E_t - E_i)/kT}\right]^2. \quad (5.9)$$

Where N_{ts} is the trap state density, N is the doping density, n_i and E_i are the intrinsic carrier density and Fermi level in the grain boundary. Under illumination Fermi level changes to quasi-Fermi level and occupancy of traps also changes. This in turn changes the barrier height. The new barrier height is given by

$$\phi_{GL} = \beta^2 \phi_{g0}, \tag{5.10}$$

where $\beta = f/f_0$, f and f_0 are the Fermi occupancy functions in the light and in the dark. The generation rate of the carriers is given by,

$$G = \sigma V_{th}\left(\frac{N_{ts}}{d}\right)\left[N \exp[-q\phi_{g0}\beta^2/kT](1-\beta f_0) - n_1\beta f_0\right]. \tag{5.11}$$

Where σ and V_{th} are the capture cross-section and thermal velocity of majority carriers. If the generation rate is known, β can be calculated using the above equation. The generation rate for air-mass number 1 (AM1) spectrum at the surface and as a function of depth have been plotted in [149]. The generation rate at the surface is close to 8×10^{21} cm^{-3} s^{-1}. The grain boundary barrier height under illumination can now be calculated. The calculated values are plotted in Fig. 1 of Ref. [149]. The change in the barrier is small for initial small decrease in the generation rate but for higher rates the barrier height falls very rapidly.

Eq. (5.9) shows the barrier height (in the dark) depends strongly on the doping concentration. The barrier height is small for low doping concentration. Initially it increases with doping concentration and obtains the maximum and with further increase of doping concentration, it starts to decrease. Most of the majority carriers go into the traps in the grain boundary below the maximum. As Fig. 5.26 shows, the carrier concentration decreases by several orders of magnitude and the conduction becomes space charge limited. We show the dark current data measured by Koliwad, and Daud [150] in poly-Si solar cells

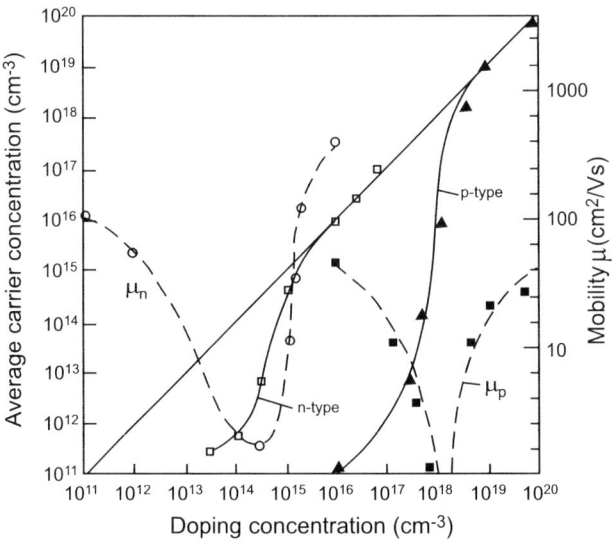

FIG. 5.26. Average free carrier concentration and mobility as a function of doping density in polycrystalline Si. The mobility values are for majority carriers. The grain size for n-Si is 25 μm and for p-Si it is 200–270 Å. The straight line is the carrier concentration versus doping concentration for single crystal silicon. Experimental points lie closely on the curves [151].

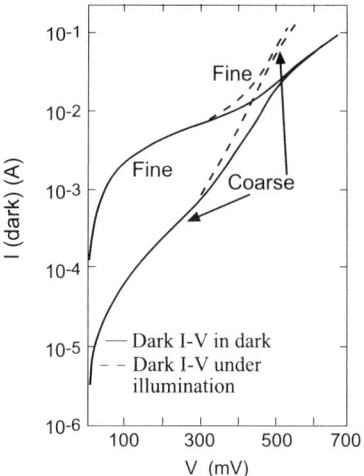

FIG. 5.27. Comparison of the measured dark current (solid line) and the dark current derived from the illuminated I–V characteristics (dashed curves) of a polysilicon solar cell. The dark currents are taken from [150]. The results for two grain sizes are shown.

in Fig. 5.27. The shapes of the currents show clearly the space charge limited nature. The short circuit currents derived by us from the illuminated characteristics are shown. The currents are higher and the shape is also changed significantly.

We now discuss the second effect. Assuming a one sided diode, the dark current is inversely proportional to the diffusion length. Using the calculated values of the diffusion length under illumination, the calculated dark current is shown in Fig. 5.28. At 50 suns the dark current decreases by a factor approximately 16. Using the equation,

$$V_{oc} = 26 \ln\left(\frac{I_{sc}}{I_d}\right) \text{MeV}, \tag{5.12}$$

the increase in the open circuit voltage turns out to be 72 MeV. It is well known that efficiency of the solar cell increases superlinearly with intensity of light. Though short circuit current increases linearly with the intensity of light, the open circuit voltage and

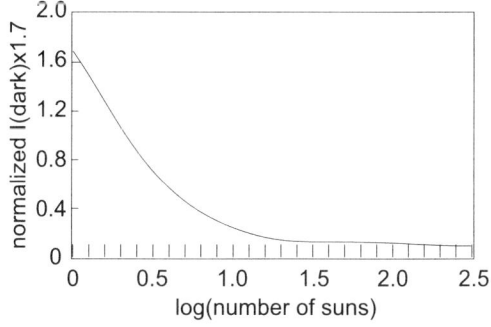

FIG. 5.28. The effect of illumination level on the dark current in polysilicon [149].

fill factor also increase. The efficiency is given by,

$$\eta = I_{sc} \times V_{oc} \times FF, \qquad (5.13)$$

the increase in efficiency becomes superlinear. In the poly-silicon solar cell there is an additional increase due to the increase of the diffusion length under the strong light. This increase is quite significant as our calculations show.

5.7. Summary and Conclusions

Diffusion length of the exciton is ten times smaller than the penetration depth of most of the light in the solar spectrum. Bilayer solar cells with a p–n junction in the active layers are not likely to give good efficiencies. Interpenetrating network of acceptor–donor polymers (or polymers and small molecules) is a very promising material for designing and manufacturing high efficiency solar cells. Transport properties of these blends are not well understood. Both experimental and theoretical work is required in this area. In a single polymer layer Schottky diode the open circuit voltage is found to be equal to the difference in work functions of the two electrodes material. However if junctions are contained in the active layer, the open circuit voltage depends on both the work function value and on the bandgap values of the constituent materials. There is an urgent need to find models, which will determine the relative contributions of the electrodes and the internal junctions to the open circuit voltage. Long term stability of the organic solar cell continues to be a problem. Better encapsulation technology is urgently required. Unpublished work of Ref. [30] on modeling of the output characteristics of the solar cells suggests that there is no space charge in a solar cell under illumination. New experiments should be designed to test this hypothesis.

CHAPTER 6
TRANSISTORS

6.1. Importance of Organic TFTs

Organic thin film transistors are fabricated with a low-temperature process. It is therefore possible to fabricate TFT arrays for flat panel displays in a low cost process. The substrates are low-cost and flexible such as polyethylene terepthalate (PET). The low cost, large area TFT arrays can be used for many applications, e.g. electronic paper, smart cards and remotely updateable posters and notice boards. Currently the amorphous-silicon-on-glass technology is used for such applications. This technology is very expensive. These applications will only become popular in marketplace if the cost of production is substantially reduced. This is the driving force for the R&D effort in organic TFTs.

Characteristics of organic TFTs reported recently are similar to those of hydrogenated amorphous silicon (a:Si:H TFTs). Field effect mobility near 1 cm^2/V s, on/off current ratio as large as 10^8, near zero threshold voltage, and subthreshold swing of less than 1 V/decade have been reported by several groups. Extensive efforts have been made on optimizing the deposition of the organic semiconductor films and improve the properties of the gate insulator to obtain better field effect mobility. With the improvement in the mobility, the effect of contact resistance becomes more severe. This effect degrades the performance most severely in the linear region. In active matrix displays TFTs are operated in the linear region.

Photoresponsivity of organic field effect transistors (photOFETs) is interesting since it is the basis for light sensitive transistors [152]. The possible applications of photOFETs are in light induced switches, light triggered amplification, detection circuits and in photOFET arrays for highly sensitive image sensors [152].

Organic semiconductors are attracting great attention for fabricating OTFTs and optoelectronic devices [153]. Organic field effect transistors (OFETs) play a prominent role in organic semiconductor based electronic devices due to the possibility of switching and driving thin film transistors (TFT) in flexible displays, smart cards, radio frequency identification (RFID) tags and large area sensor arrays. Results of the performance of organic semiconductor devices are now comparable with the amorphous Si devices [154]. The performance of optimized pentacene-based organic thin film transistors (OTFTs) has become so good that it is now possible that OTFTs are used as drivers of organic light emitting diodes (OLEDs) in high resolution active matrix displays. As emphasized by Angelis et al. [154], it is very important to understand the transport mechanisms in the OTFT structure. In fact, many problems including ageing effects can be solved if transport mechanism is well understood.

Several π-conjugated polymers and oligomers (polyacetylene, several polythiophenes, Ooct-OPV5 oligomers, and phthalocyanines) have been used for the active layer of the transistors.

6.2. Early Work

Francis Garnier [3, and references given therein] fabricated the first transistor using molecules of sexithiophene. The transistor could be twisted, bent or rolled without degrading its characteristics. Computers fabricated using these devices will work at less than one thousandth of the speed of those made with amorphous Si transistors. They would be useful in video displays and liquid-crystal displays. In active matrix displays, each pixel is controlled individually by a thin film transistor. A 50 cm full color display contains more than two million pixels. Organic transistors, considerably cheaper than the amorphous Si transistors being used at present, will be a boon to the manufacturers.

A schematic diagram of an early thin film transistor fabricated in 1995 [155] using α-6T polymer for the active layer is shown in Fig. 6.1. The field conductance of the same transistor is shown in Fig. 6.2. A review of the early work has been published by Horowitz [156]. Measurements of characteristics of Ooct-OPV5 (5-ring n-octyloxy-substituted Oligo[p-phenylene vinylene]) thin film transistor were made by Jain et al. Ooct-OPV5 was dissolved in chloroform and spin-cast into thin films. The transistor showed good I_{ds}–V_{ds} curves for different values of gate voltage. The saturation of the gate current occurred at a gate voltage of ~-40 V for all values of gate voltage. The ON/OFF ratio was $\sim 10^6$. The mobility derived from the analysis of these characteristics is $\mu_{FE} = 2.56 \times 10^{-4}$ cm^2/V s.

Jain et al. [127] fabricated and characterized an oligomer OFET. A SiO$_2$ layer was first deposited on a Si substrate. An interdigitating Au-finger structure was formed using a lift-off technique. Ooct-OPV5 oligomer with polystyrene (PS) active layer was deposited over this structure. The square root of the measured drain current I_d is plotted as a function of gate voltage in Fig. 6.3. To interpret the experimental results, theory of Si FETs is generally used. At high drain bias the current saturates and becomes independent of drain voltage. The saturation drain current is given by,

$$I_{d,sat} = \frac{\mu_{FE} W C_{ox}}{2L}(V_g - V_t)^2. \quad (6.1)$$

The subscript FE in the mobility μ_{FE} is used to emphasize the fact that the mobility determined from the OFET data is different from that determined from the diode data. W is the device width, C_{ox} is the capacitance of the oxide layer, and V_t is the threshold voltage.

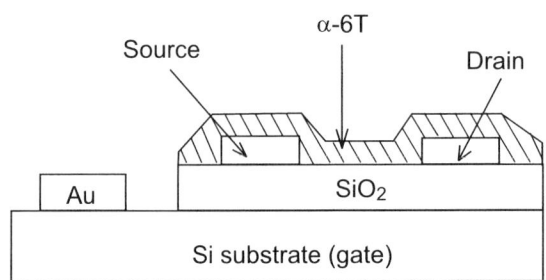

FIG. 6.1. Schematic diagram of the thin film transistor using an α-6T polymer for the active layer [155]. Gold source and drain pads are fabricated using the photolithographic technique. The gate length varied between 2.5 and 150 nm and the width was 250 µm. The polymer (from 2.5 to 150 nm thick) is sublimed over the contacts.

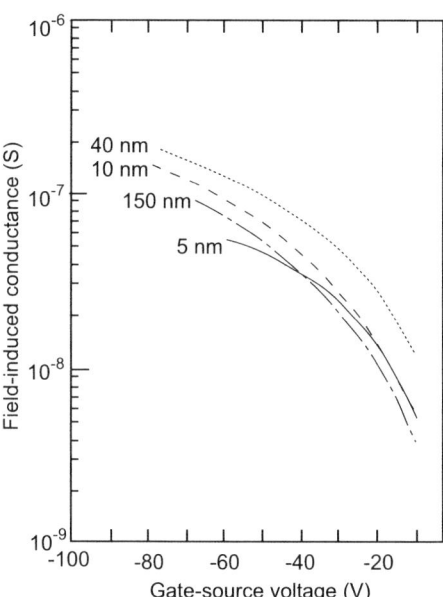

FIG. 6.2. Field-induced conductance of the α-6T thin film transistor as a function of gate source voltage for different thicknesses. The channel length of this transistor is 12 μm and the drain source voltage is -100 V [155].

In practice large gate voltages are used and threshold voltage is small. Therefore according to this equation a plot of $\sqrt{I_{d,sat}}$ versus V_g should be a straight line with its slope proportional to μ_{FE}. This provides a method for determining the field effect mobility of the material of active layer. The mobility in Ooct-OPV5:PS measured in this manner is 3.24×10^{-4} cm^2/V s.

A comparison of the polymers and amorphous inorganic semiconductors is interesting. The conductivity is small in both cases due to the presence of large concentration of traps. The plot of ln I versus $1/T$ has many similar features. In both cases the transport

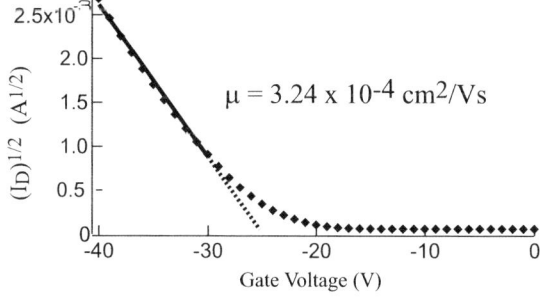

FIG. 6.3. $\sqrt{I_d}$ versus gate voltage in saturation regime of the organic FET. The active layer consisted of Ooct-OPV5:PS blend. The drain source voltage was -30 V [127].

is dominated by drift of carriers in the extended states at high (near room) temperature and by tunneling and hopping mobility at low temperatures. However there is one important difference between the two materials [157]. In amorphous Si the atoms are bound by covalent interactions. In the organic compounds the atoms are weakly bound by van der Waals forces. Consequently hydrogenated amorphous Si has wide conduction and valence bands, and the traps are distributed over the whole bandgap. The conduction and valence bands in organic semiconductors are narrow. The trap distribution, at least in some polymers, is confined to a narrow energy range, e.g. in alpha-conjugated sexithiophene (α-6T) [157].

6.3. Effect of Traps

Horowitz and Delannoy [157] have modified the TFT theory to take into account the presence of the traps and the trapped charges. The main equations used in their model are given below.

$$n_f = \frac{N_c}{1 + \exp[(E_c - E_F - qV(x))/kT]}, \quad (6.2)$$

$$n_t = \frac{N_t}{1 + \exp[(E_t - E_F - qV(x))/kT]}, \quad (6.3)$$

$$F_x^2(V) = -\frac{2}{\epsilon\epsilon_0} \int_0^{V(x)} \rho_{tot}\, dV, \quad (6.4)$$

$$Q_{st} = \int_0^{V_s} \frac{\rho_t}{F_x(V)}\, dV, \quad (6.5)$$

$$Q_{sf} = \int_0^{V_s} \frac{\rho_f}{F_x(V)}\, dV, \quad (6.6)$$

$$Q_{total} = Q_{st} + Q_{sf}. \quad (6.7)$$

The symbols used in the above equations are: n_f and n_t are densities of free and trapped charge carriers in the volume of the active layer, T is the temperature, N_t is the trap density and N_c is the conduction band density of states. The axes are defined in Fig. 6.4(a). E_c is the conduction band edge, E_F is the Fermi level and E_t is the trap depth (see Fig. 6.4(b)). F_x is electric field along the x-direction, Q_{st} and Q_{sf} are the trapped and free surface charge densities, and $\rho = qn$ is the volume charge density. Note that the currents in the transistor are not space charge limited. Space charge which limits the current (in polymer diodes for example) is created by the rapidly varying charge carrier profile in the direction of the current flow. The situation in the FET is quite different. The carriers are induced by the gate voltage uniformly from source and drain.

The difference between the above theory and the conventional theory used for polymers is the inclusion of Eqs. (6.3) and (6.5). When traps are introduced the total charge in the active layer increases and therefore surface potential and gate voltage also change. If we

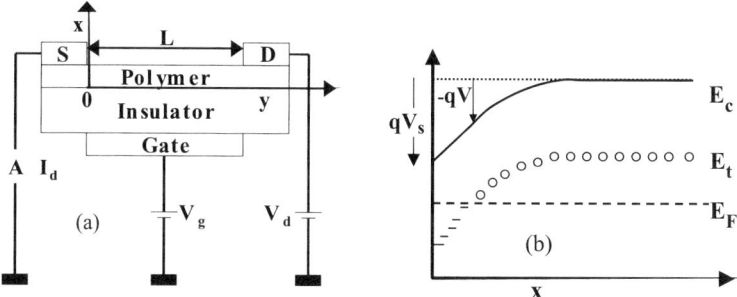

FIG. 6.4. (a) Schematic diagram of an n-type organic thin film transistor. (b) Energy level scheme of an accumulation layer for an n-type semiconductor with a single trap level of energy E_t. qV_s is the band bending at the surface [158].

hold the surface potential at a fixed value, the free charge densities will change when the traps are introduced. Numerical results of Horowitz and Delannoy [157] demonstrate that the characteristics of the OFET are considerably modified by the presence of traps.

The electric field at the interface of the oxide and the active layer is large. As discussed earlier, the field assists the ionization of the traps due to the Poole–Frenkel Effect. The trap depth is reduced by an amount $\beta\sqrt{F}$ and the number of trapped carriers is reduced. Following Horowitz and Delannoy [157] we consider the traps at a single level located near the band edge [158] in an n-type polymer. The treatment is quite general and can be extended to p-type polymers quite easily. When PFE is included, Eq. (6.3) changes to,

$$n_t = \frac{N_t}{1 + \exp[(E_t - E_F - qV)/kT]\exp[(-\beta_{PF}\sqrt{F(V)})/kT]}, \quad (6.8)$$

where

$$\beta_{PF} = \left(\frac{q^3}{\pi\epsilon\epsilon_0}\right)^{1/2}. \quad (6.9)$$

The effect of screening on β can also be taken into account, see [158, and references given therein] for details. However the theory can be fitted with the experiment without the screening effect. Introduction of Eq. (6.8) makes the calculation of the trapped carrier density n_t difficult. This equation involves the electric field F and Eq. (6.4) shows that the calculation of electric field involves both free and trapped carrier densities. A self-consistent numerical calculation of electric field and trapped carriers is made using the method of iteration. First electric field $F(V)$ is calculated using Eqs. (6.2), (6.3), and (6.4) and neglecting field dependent trap occupancy (FDTO). Using this electric field a new value of n_t is then calculated using Eq. (6.4). Electric field is again calculated using this new value of n_t. This process of iteration is repeated until reproducible values of n_t and F are obtained.

6.4. High Field Effects

A schematic diagram of an organic thin film transistor (OTFT) fabricated by Rashmi [158] is shown in Fig. 6.4(a). Here L is the channel length and S and D are the source

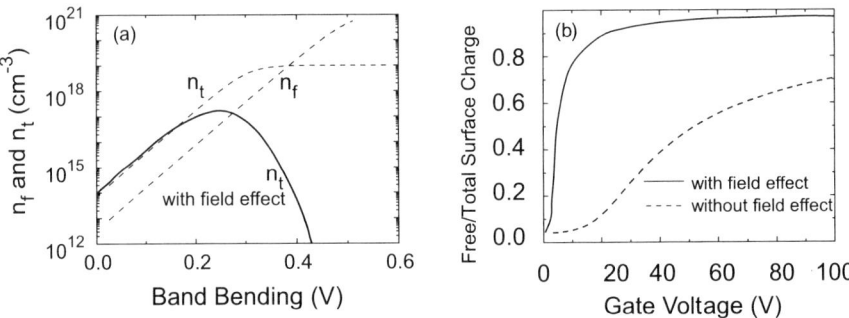

FIG. 6.5. (a) Trapped carrier density n_t (with and without field effect) and free carrier density n_f as a function of band bending in a semiconductor with one single level trap at energy 0.2 eV from conduction band and $N_t = 10^{19}/\text{cm}^3$. Other parameters are listed in column A of Table 6.1. (b) Free surface charge density to total induced charge ratio versus gate voltage in an n-type organic TFT with a trap level at 0.2 eV from conduction band edge. Other parameters are listed in column A of Table 6.1 [158].

and drain. When a voltage is applied to the gate, band bending takes place in the active layer as shown in Fig. 6.4(b). Several other papers were published on fabrication and characteristics of OFETs in the early years [159,157,158, and references given therein]. The quantity of practical interest is the dependence of carrier densities on the gate voltage (and not only on the band bending or surface voltage). The gate voltage V_g and surface voltage V_s are related by the properties of the gate insulator,

$$V_g = V_i + V_s, \tag{6.10}$$

where V_i, the voltage drop across the insulator, is given by,

$$V_i = \frac{\epsilon \epsilon_0 F_s}{C_i}. \tag{6.11}$$

The calculated free and trapped carrier densities are shown by the dashed curves in Fig. 6.5(a). The trapped carrier density is much larger than the free carrier density at the edge of the active layer away from the gate where the band bending voltage V is small. As one moves closer to the gate and to larger band bending, the trapped carrier density saturates and the free carrier density becomes larger than the trapped carrier density. This result is similar to that obtained with polymer diodes and shown in Fig. 3.8.

The field is a function of band bending voltage V and since V depends on x, field is also a function of x. We first calculate the charge distribution and the field F_x as a function of the surface potential V_s. The gate voltage is then calculated by adding the drop across the insulator to V_s. To calculate the drain current we use the following equations [157],

$$I_d = \frac{W}{L}\mu \int_{V_g-V_d}^{V_g} |Q_{sf}(V_g)| \, dV_g, \quad V_d < V_g, \tag{6.12}$$

and

$$I_{d,\text{sat}} = \frac{W}{L}\mu \int_0^{V_g} |Q_{sf}(V_g)| \, dV_g, \quad V_d > V_g. \tag{6.13}$$

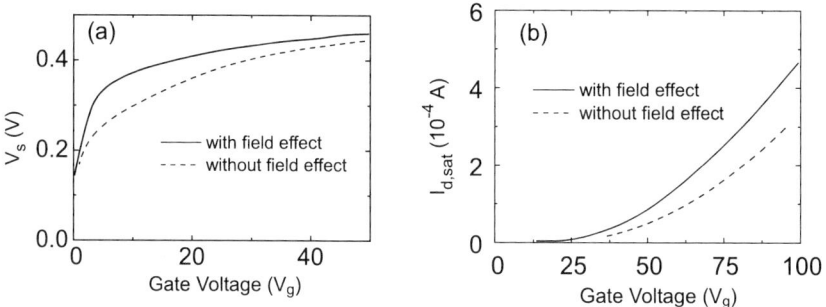

FIG. 6.6. (a) Calculated values of surface potential V_s plotted as a function of gate voltage. The dashed curve is obtained using the theory of Horowitz and Delannoy [157]. The solid curve is the result based on the FDTO model. (b) Saturation drain current ($I_{d,sat}$) versus gate voltage (V_g) plot of the numerically computed current of an Organic TFT (with and without the field effect). Parameters used are given in column A of Table 6.1 [158].

$I_{d,sat}$ is the saturation current. A key parameter is the ratio θ_s of the free surface charge density to total induced charge density [157]. θ_s is plotted as a function of gate voltage in Fig. 6.5(b).

The effect of FDTO on surface potential is shown in Fig. 6.6(a). In this figure surface potential is plotted as a function of gate voltage. For a given gate voltage, surface potential increases in the FDTO model. At $V_g \sim 10$ V band bending increases by about 25%. The calculated $I_{d,sat}$ is shown in Fig. 6.6(b). The saturation current increases faster when FDTO is used. The increase is very large at high gate voltages.

The above discussion demonstrates that high field effects included in the FDTO model modify the OFET characteristics very significantly. The FDTO model is in very good agreement with the experimental results. The model has been compared with two sets of experiments [158]. The experimental data for α-6T [157] is compared with the FDTO

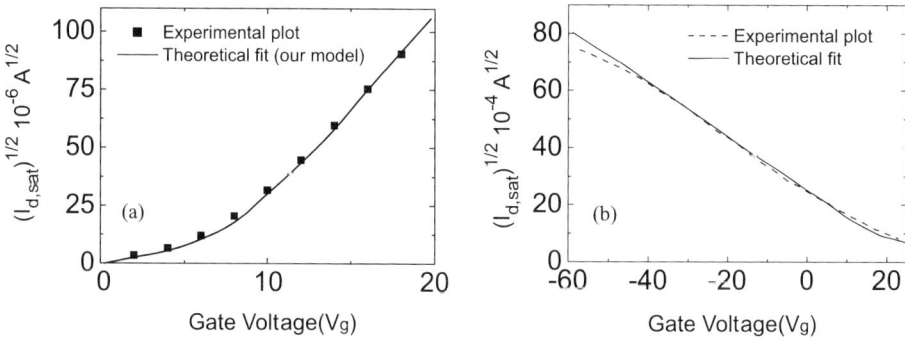

FIG. 6.7. (a) Numerically computed $\sqrt{(I_{d,sat})}$ versus V_g plot compared with experimental curve of an α-6T TFT obtained by Horowitz and Delannoy [157]. The calculations have been made using the FDTO model. (b) Same as (a) but for a pentacene TFT. The calculated values are shown by the solid curve. The dashed curve represents experimental data taken from Klauk et al. [160]. The values of the parameters used in the fitting the data with the model are given in column B for (a) and in column C for (b) of Table 6.1. The figure is taken from [158].

TABLE 6.1
VALUES OF THE PARAMETERS USED IN THE NUMERICAL CALCULATIONS OF FIGS. 6.5–6.7 [158]

Nomenclature		A	B	C
Temperature (K)	T	300	300	300
Density of states (per cm^3)	N_c	10^{21}	–	–
Density of states (per cm^3)	N_v	–	5×10^{21}	5×10^{21}
Trap density (per cm^3)	N_t	10^{19}	2.5×10^{19}	2×10^{19}
Trap level (eV)	$E_t - E_v$	–	0.23	0.15
Fermi level (eV)	$E_c - E_F$	0.5	–	–
Fermi level (eV)	$E_F - E_v$	–	0.5	0.5
Mobility (cm^2/V s)	μ	0.1	0.0013	0.2
Dielectric constant	ϵ/ϵ_0	2	2	2
Insulator capacitance (nF/cm^2)	C_i	10	25	17
Gate width (cm)	W	0.1	0.031	0.005
Gate length (cm)	L	0.001	0.009	0.001

The values used in fitting the theory with the experimental data are given in column A for Figs. 6.5–6.6, in column B for Fig. 6.7(a) and in column C for Fig. 6.7(b). Horowitz and Delannoy [157] have used $N_t = 4.2 \times 10^{18}$/cm^3 and $N_v = 1.7 \times 10^{21}$/cm^3.

model in Fig. 6.7(a). Pentacene TFT experimental data taken from Klauk et al. [160] is compared with the model in Fig. 6.7(b). The agreement is very good in both cases.

Numerical calculations show that E_t affects significantly the $\sqrt{I_d}$ versus V_g plots. The other parameters have smaller effect on the drain current. Some of the parameters obtained with the FDTO model are different from those obtained in Ref. [157]. To obtain accurate values of the material parameters, the use of a model based on sound physics is essential.

6.5. Transport in Polycrystalline Organics

6.5.1. Effect of Grain Boundaries

It is known for many years that the mobility of highly pure and defect free crystals of small molecules follow the same behavior as the inorganic semiconductors [161]. The mobility decreases with temperature. This indicates that transport is in the bands. However small molecular polaron models can also account for this temperature behavior. Horowitz [161] has investigated the current flow in thin film Organic Field Effect Transistors (OFETs) based on polycrystalline organic materials. Horowitz [161] developed a model in which transport takes place in the delocalized states but it is restricted by the distribution of shallow traps near the band edges. The polycrystalline materials consist of two regions: grain which are more like a single crystal and the grain boundary which contain large densities of defects and traps. The modeling of two dimensional case is necessary for thin films. The modeling of the two dimensional is different from the three dimensional case. It is not possible to obtain an analytical solution for the problem. The problem is simplified if one assumes that the component of the electric field along the channels is much smaller than across it. This is the well-known graded channel approximation. The field across the channel is given by Gauss's law. Horowitz [161] also

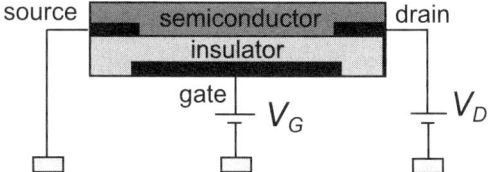

FIG. 6.8. Schematic view of the transistor used by Horowitz [161].

fabricated a OTFT to validate his model. The structure fabricated by Horowitz is shown in Fig. 6.8. Sexithiophene organic was used for active layer. The structure of sexithiophene is shown in Fig. 6.9. The Rod-like molecules stack in layers, long axes are parallel to each other. The thickness of the layer DN is somewhat smaller than the length of the rod-like molecules (see Fig. 6.9).

To simplify the problem, Horowitz assumed that the traps are at a single energy level. Barrier height is the maximum at the gate voltage more than about 30 V. The longitudinal field applies only in the space charge region which does not extend far from the GB. The variation of W was calculated. Except at very low gate voltage, W is not sensitive to the temperature. W becomes less than 1 nm at a gate voltage more than 20 V.

The numerical analysis showed that the transport is limited by a barrier created at the grain boundaries (GBs). A low doping in the material was considered in the analysis. However, a large density of free carriers is created by a voltage applied to the gate. The gradual-channel approximation was used. The width of the barrier was found to be smaller than the width of the GB. The transport of charge carrier was modeled based on tunneling through the barrier. The tunneling was assumed to be elastic, the barrier was assumed to be square in shape and one electron approximation was used. To check the model, gate voltage dependent mobility was measured at 11 K on a sexithiophene transistor. The measured value agreed with the results obtained by numerical calculations for reasonable values of trapped carrier density and GB width. However, the calculated value of mobility was much higher than the measured one. The probable cause for this discrepancy is that actual tunneling is not elastic.

6.6. Pentacene TFTs

The OTFT parameters, field effect mobility, modulated on/off current ratio, and threshold voltage, depend on the molecular structure of the semiconducting film. Molecular ordering and crystalline orientation of the organic semiconductor in the thin film have a large affect on the performance of the TFTs. The charge carriers electrons or holes generally increase with the molecular chain ordering or improved crystalline quality.

Aromatic hydrocarbon pentacene is important for application in OTFTs as it has superior field effect mobility, good semiconducting behavior, and stability [153]. The chemical structure is shown in Fig. 6.10. Pentacene semiconductor films can be fabricated by sublimation in a vacuum deposition system. Optimization of the fabrication parameters, such as the substrate temperature and deposition rate, can yield a highly ordered pentacene film with improved device performance. Oriented films have optical and electrical anisotropies.

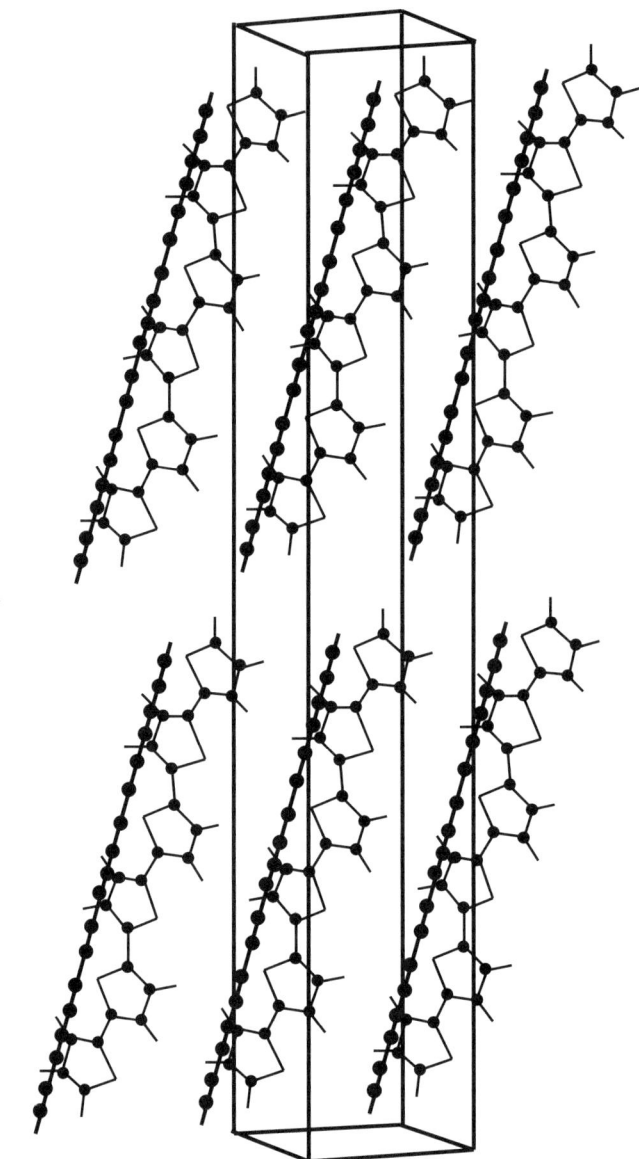

FIG. 6.9. Structure of sexithiophene [161].

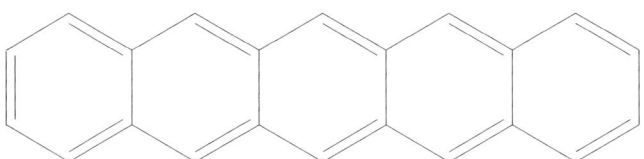

FIG. 6.10. Schematic of the chemical structure of pentacene.

The organic TFT arrays must demonstrate a level of performance at least equal to the amorphous silicon state of the art before they can be widely used. The only material that has shown such performance is pentacene. Angelis et al. [154] fabricated high quality polycrystalline pentacene-TFTs and measured the electrical characteristics at different temperatures. They obtained high quality transistors by using polymethylmetacrylate (PMMA) buffer layers. The technique of improving the quality of these TFTs by using PMMA buffer layers is well known. Angelis et al. [154] determined the effective density of states of the localized states and the carrier mobility. Devices were made in two configurations, with TOP contact (TC) and BOTTOM contact (BC). Heavily doped silicon wafers were used as substrates; they acted as gate electrodes also. The gate insulator was formed by thermally oxidizing Si. (Silicon dioxide thickness was $d_{ox} = 150$ nm.) In BC configuration the source and drain gold contacts (30 nm thick) were defined by optical lithography and wet etching. The channel lengths $L = 7, 15, 25, 100, 500$ μm and channel width $W = 200$ μm were used. Thin buffer PMMA layers were deposited on the substrate. The thin PMMA films improve the quality of the polycrystalline pentacene. Thermal evaporation of 97% purity pentacene was used to fabricate the active layers. For top contact (TC) configuration, source/drain gold contacts were evaporated through a shadow mask on top of pentacene active layer. The length and width of the contact layers were: $L = 100$ μm, $W = 200$ μm.

Transfer characteristics of both top and bottom configurations of the pentacene devices were measured at $V_{ds} = -1$ V. The field effect mobility in TC devices was $\mu_{FE} = 1.4$ cm^2/V s. The values of subthreshold slope and threshold voltage were $S = 0.5$ V/dec and $V_{th} = -13.5$ V. BC devices ($L = 100$ μm) show $\mu_{FE} = 1.1$ cm^2/V s, $S = 0.3$ V/dec, and $V_{th} = -6.6$ V. At the time the paper was published, the performance was best for long-channel pentacene-based BC TFTs. The output characteristics of TC and BC devices are shown in Fig. 6.11. They show a good linear behavior at low V_{ds} as well as an excellent saturation region at high V_{ds}. There is a difference between the characteristics of the TC and BC devices. In TC device the current flows from the channel to the drain through the pentacene active layer whereas in the BC configuration holes are injected from the gold source contact into the pentacene channel through the PMMA buffer layer. They are extracted from the drain through the PMMA. The buffer layer introduces a parasitic resistance and this reduces the μ_{FE} in the BC devices. With shorter channel lengths $\mu_{FE} = 0.37$ cm^2/V s has been obtained for $L = 7$ μm. However, the thin PMMA buffer layer still allows sufficiently good ohmic contacts.

Transfer characteristics of TC pentacene TFTs have also been measured at different temperatures in the range between 205 and 300 K (see Fig. 6.12). As shown in Fig. 6.13, the field effect mobility and threshold voltage, evaluated from the slope of the linear portion of the transfer characteristics, monotonically increase with temperature.

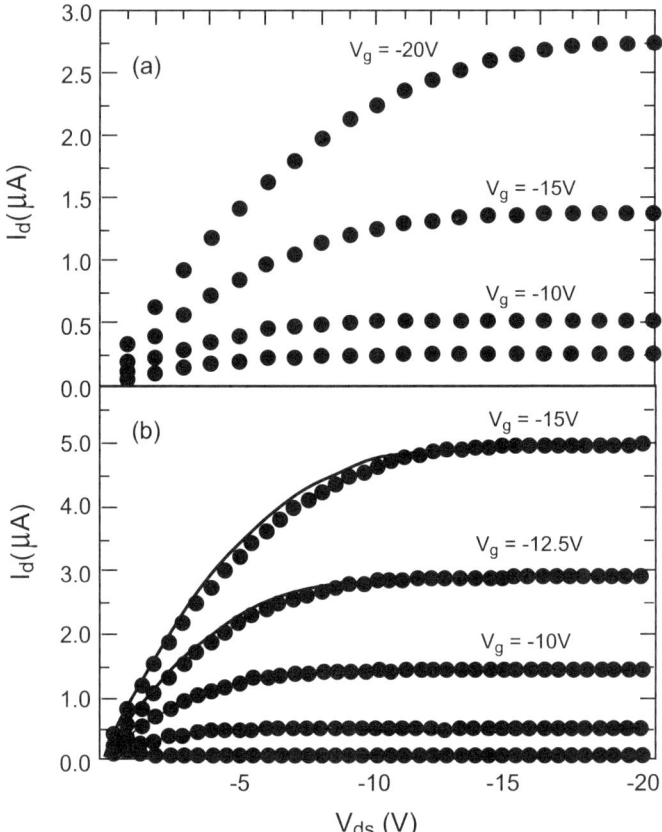

FIG. 6.11. Experimental (symbols) output characteristics of (a) BC and (b) TC pentacene TFTs, measured at $T = 300$ K and different gate voltages ($L = 100$ μm, $W = 200$ μm, $d_{ox} = 60$ nm). Also the solid lines are the simulated output characteristics of TC pentacene TFT [154].

The DOS was extracted from the analysis of the sheet conductance, G, at different temperatures. The calculated DOS (see Fig. 6 of [154]) can be represented by the sum of two exponential tails:

$$N(E) = N_t \exp\left(\frac{E - E_F}{E_t}\right) + N_d \exp\left(\frac{E - E_F}{E_d}\right), \tag{6.14}$$

where $N_t = 1.2 \times 10^{21}/\text{cm}^3$ eV, $E_t = 17$ meV, $N_d = 6 \times 10^{19}/\text{cm}^3$ eV and $E_d = 85$ meV. This approximated DOS was used in the 2D numerical device analysis program DESSIS, using conventional drift-diffusion transport model, to simulate the transfer and output characteristics at different temperatures. Figs. 6.11 and 6.12 show that the transfer and output characteristics are very well reproduced in the temperature range considered. The model based on the spatially uniform DOS and conventional drift-diffusion transport model is quite useful for analyzing the transfer characteristics of pentacene TFTs and that the DOS, obtained from this procedure is quite accurate.

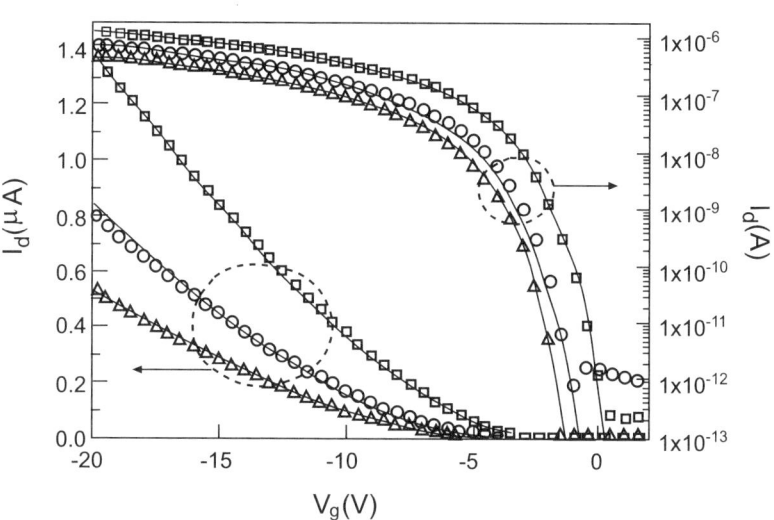

FIG. 6.12. Simulated (lines) and experimental (symbols) transfer characteristics (measured in vacuum at $V_{ds} \approx -1$ V), of as fabricated TC pentacene TFTs for three different temperatures: 300 K (squares), 240 K (circles), 205 K (triangles) ($L = 100$ μm, $W = 200$ μm, $d_{ox} = 60$ nm) [154].

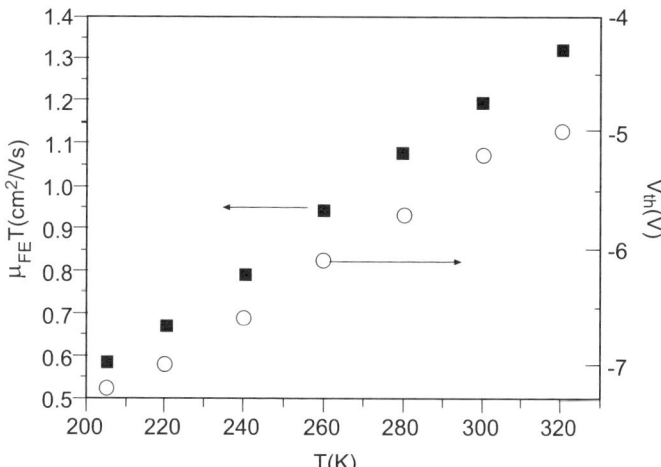

FIG. 6.13. Field effect mobility, and threshold voltage V_{th} versus temperature of TC pentacene TFT [154].

To summarize this work, Angelis et al. [154] fabricated high quality pentacene OTFTs in two configurations, i.e. with top and bottom contacts. They determined a field effect mobility higher than 1 cm^2/V s. The transfer characteristics of the top contact TFTs were measured at different temperature and analyzed using a model based on the assumption of a spatially uniform distribution of localized states. The calculated DOS can be represented approximately by the sum of two exponential tails given by Eq. (6.14). The calculated DOS is an effective DOS, including the contributions of the in-grain and grain boundary defects as well as of the dielectric/pentacene interface defects. Using this DOS, transistor characteristics were calculated using the 2D numerical device analysis program DESSIS, assuming a drift-diffusion transport model.

Anisotropic mobility is useful in isolating neighboring components to reduce the crosstalk in logic circuits or pixel switching elements used in displays. Recently, several methods have been developed to align organic semiconducting films with anisotropic mobility, including mechanical stretching, liquid crystalline self-organization, rubbing, and photoalignment. The ion-beam (IB) alignment method, a non-contact alignment technique, is applied in [153] to align pentacene molecules. The IB alignment method can generate an easy-orientation axis on the surface of a polymer (polyimide). The alternative rubbing technique has drawbacks, e.g. sample contamination, static charge generation, production of scratches. Chou et al. [153] have used successfully ion-beam alignment and used it to generate pentacene OTFTs with anisotropic electrical characteristics and carrier mobility that was an order of magnitude higher than that of a typical device with pentacene grown directly on the surface of native SiO$_2$ (see Table 6.2 and Fig. 6.14). The thin films of pentacene were studied by X-ray diffraction, scanning electron microscopy, polarized photoluminescence, and polarized Raman spectroscopy. The authors investigated the anisotropy if the OTFTs fabricated with IB alignment technology. Interpretation of Raman spectroscopy measurements showed that IB alignment technology increased carrier mobility within an active layer. The microstructure of pentacene films was studied using polarized photoluminescence and Raman spectroscopy.

Mottaghi and Horowitz [164] have investigated the degradation of carrier mobility in pentacene TFTs due to effect of electric field. They fabricated pentacene-based OTFTs with bottom-gate, top-contact architecture. Alumina was used as substrate. In one set of devices pentacene was deposited directly on alumina. In the second set of devices a fatty

TABLE 6.2
THE EFFECT OF THE ION BEAM TECHNOLOGY ON THE PENTACENE FILMS [153]

	Ion-beam direction with respect to current flow	Crystal size (Å)	Crystal disorder (%)	Mobility of linear region (cm^2/V s)	Mobility of saturation region (cm^2/V s)	Dichroic ratio (linear region)	Dichroic ratio (saturation region)
Ion bean Aligned	perpendicular	287.2	1.08	0.141	0.155	2.39	2.19
	parallel			0.059	0.0629		
Nonaligned	non	230.9	1.38	0.005	0.019		

Crystal size, crystal disorder, and field effect mobility of pentacene films grown on the surface of ion-beam treated and native SiO$_2$. Dichroic ratio signifies alignment of pentacene perpendicular to the ion-beam direction.

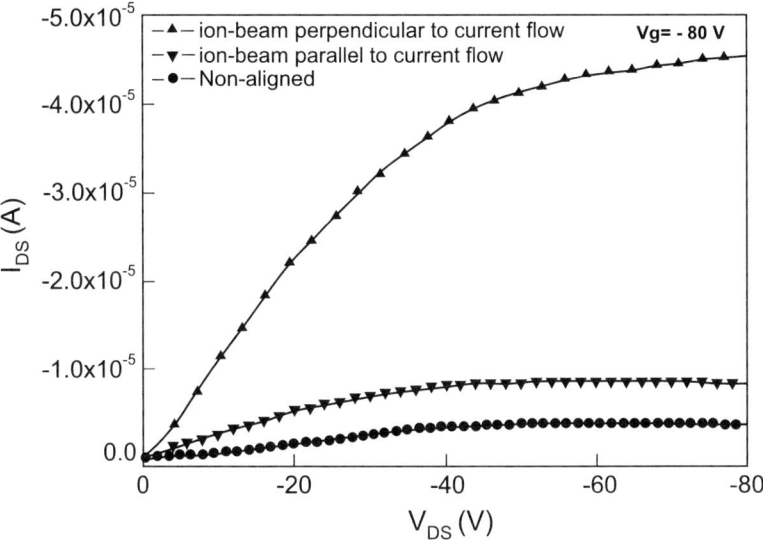

FIG. 6.14. Drain source current drain source voltage characteristics of three pentacene OTFTs. The up and down triangles indicate the data from devices with ion-beam treated SiO_2 layers, in which the treated directions of the ion beam are perpendicular and parallel to the direction of the current flow in the channel, respectively, while the circles represent the data from the device without the ion-beam processing [153].

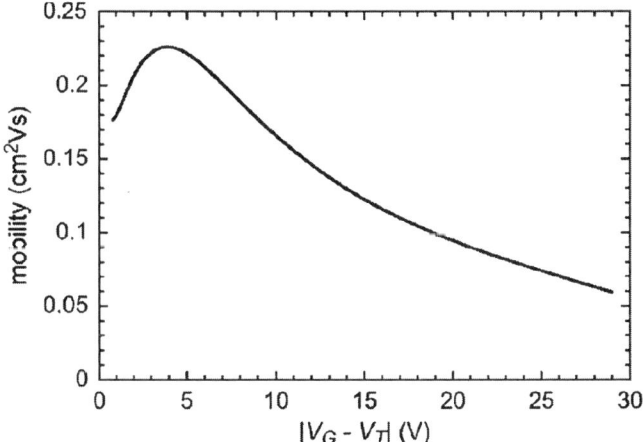

FIG. 6.15. Gate voltage dependent mobility of the device [164].

acid self-assembled monolayer (SAM) was first deposited on alumina and pentacene was deposited on the monolayer. The use of SAM buffer decreases substantially the grain size in the polycrystalline films. Mottaghi and Horowitz [164] analyzed carefully their experimental data and determined the threshold voltage, and contact resistance. They then determined the gate-voltage dependent mobility in both sets of devices. The mobility is found to first increase at low bias, and then decrease at higher gate voltage as shown in Fig. 6.15. They could explain the latter behavior through an estimation of the distribution of charges across the accumulation layer. They found good agreement if it is assumed that the mobility in the layer next to the insulator was negligible as compared to that in the bulk of the film. The initial rise of the mobility is interpreted in terms of multiple trap and release (MTR) with a distribution of traps located in the grain boundaries. The mobility was corrected for both contact resistance and mobility degradation. The corrected mobility was around 3 $cm^2/V\,s$ for pentacene deposited on bare alumina, and 5 $cm^2/V\,s$ when pentacene is deposited on SAM modified alumina. The authors suggested that the smaller grains grown on modified alumina are more regular, and hence less defective, than the larger grains deposited on bare alumina.

Guo et al. [165] prepared the pentacene films by the organic molecular beam deposition method. They investigated the effect of thermal annealing on the morphology and carrier mobility of the pentacene thin film transistors (TFTs). The effect on morphology is shown in Fig. 6.16. For all the TFT samples the plots of mobility versus $1/T$ were Arrhenius as shown in Fig. 6.17. The mobility is controlled by a thermally activated transport that could be explained by the carrier trap and thermal release transport mechanism. In order to investigate the annealing effect, they tested the data for a significant period of time after annealing until the temperature recovered to room temperature, so that the thermal activation effect was screened and possible effects of thermal expansion and stress were also ruled out. Guo et al. [165] showed that when thermal activation effects are eliminated, the mobility increased at annealing of a low temperature of 45 °C. If the annealing is done at temperature >50 °C the mobility decreased. Structural studies showed that structure changes take place at a temperature much lower than the melting point. The grain size decreased and XRD peak intensity decreased. These changes take place at the lower temperature because of the weak binding due to van der Waals forces.

6.7. Contacts

Ohmic contacts in Silicon metal-oxide-semiconductor field effect transistors (MOSFETs), polycrystalline Si TFTs, and a:Si:H TFTs are made by using heavily doped regions. Attempts were also made to selectively dope the contact regions of OTFTs by intercalating electron accepting molecules, such as $FeCl_3$ or I_2. The results were not very encouraging. Since then, only small amount of work has been done on "doping" methods for forming contacts. Now Schottky source and drain contacts are generally formed. In this work effect of the device processing, choice of contact metal, and device design have not been discussed. Gundlach et al. [163] have investigated the parasitic contact effects in pentacene TFTs. They fabricated the transistors using different designs and contact metals. Pentacene TFTs with gold (Au) contacts extracted from top contact (TC) and

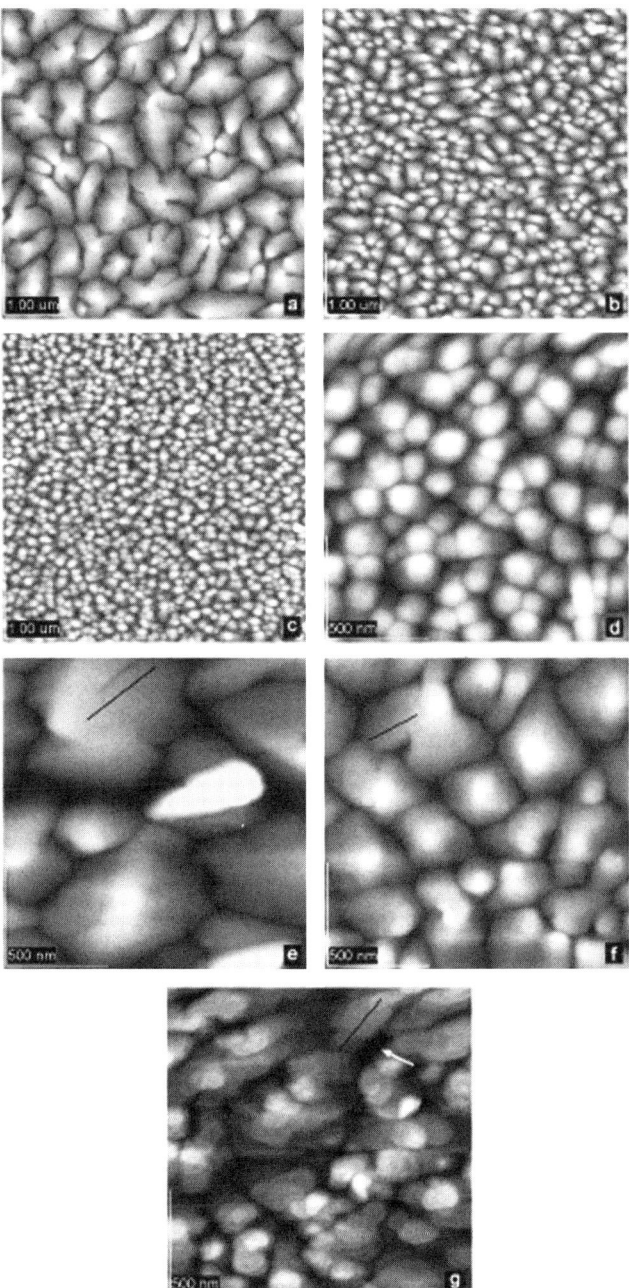

FIG. 6.16. AFM images showing the morphology of (a) a room temperature (RT) reference sample without annealing, (b) the 45 °C annealed sample, (c) the 50 °C annealed sample; (d) magnified image of the 50 °C annealed sample, (e) magnified image of the RT sample, (f) magnified image of the 45 °C annealed sample, and (g) magnified image of the 70 °C annealed sample, where the white arrow indicates a cavity showing the bare substrate. The corresponding height profiles of the black lines indicated in (e)–(g) are also illustrated [165].

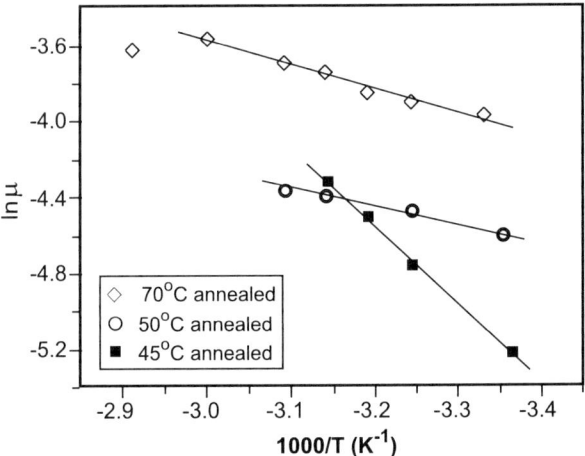

FIG. 6.17. Arrhenius plot of the logarithm mobility versus the reciprocal temperature [165].

bottom contact (BC) gated-TLM test structures, were studied. BC pentacene TFTs with palladium (Pd) contacts were also investigated. Choice of contact metal affects the charge injection of charge carriers very significantly. The device operation and performance depend strongly on the device design and processing. Energy-band diagrams are of limited use for selecting contact metals. If the source and drain contacts are of asymmetric metals the charge injection at the source contact is more important for the extrinsic device performance (see Fig. 6.18). Importantly, Gundlach et al. [163] showed that parameters used to describe the device performance and operation, such as V_t and μ, are strongly dependent on the contacts and may not reflect the channel properties of the device. The parasitic resistance R_p in TC and BC pentacene TFTs was determined to be similar to R_{ch} and scaling studies show that OTFT performance may be contact limited for $L < 10$ μm. Preliminary 2D simulations show that BC TFT performance is more strongly affected by the formation of a Schottky barrier than TC TFT performance. Some other effects, e.g. tunneling mechanisms, film morphology and microstructure, contact contamination and chemistry, and in-gap trap states were not taken into account. The results for BC devices with Pd contacts indicate that the film microstructure at the contacts affects the charge injection.

6.8. Organic Phototransistor

Several organics, e.g. pristine poly(3-octylthiophene), polyfluorene, bifunctional spiro compounds and polyphenyleneethynylene derivative, have been used for fabricating photOFETs. Responsivity as high as 0.5–1 A/W has been achieved in some of these transistors. We have already discussed the bulk heterojunction concept in Chapter 5. The bulk heterojunctions are fabricated using acceptor materials with high electron affinity (such as C_{60} or soluble derivatives of C_{60}) mixed with conjugated polymers as electron donors. PhotOFETs based on conjugated polymer/fullerene blends are expected to show

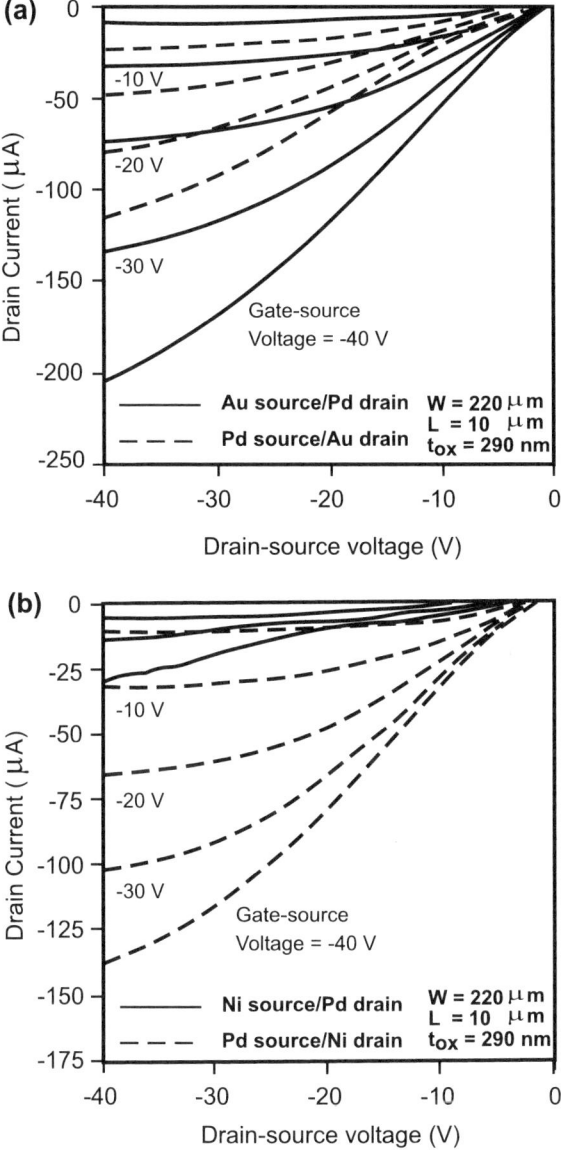

FIG. 6.18. Output characteristics for BC pentacene TFTs with (a) Au–Pd and (b) Ni–Pd contacts. All TFTs had dimensions of $L = 10$ μm, $W = 220$ μm, and SiO$_2$ gate insulator thickness of 290 nm. The dashed lines correspond to the devices biased with the Pd contact as the source [163].

higher photoresponsivity as compared to devices with single components. Marjanovic et al. [152] fabricated photoresponsive organic field effect transistors (photOFETs) based on conjugated polymer/fullerene solid state blends as active semiconductor layer. The polyvinyl-alcohol (PVA) or alternatively divinyltetramethyldisiloxane-bis(benzocyclobutene) (BCB) as gate dielectrics. When PVA is used as dielectric, the responsivity of the transistor is high but photostability is poor. If photOFETs fabricated with BCB as dielectric the transistor behavior in a broad range of illumination intensities is good and photostability is also good even at high illumination.

The chemical structure of PCBM and MDMO-PPV was shown in Fig. 5.15. The chemical structure of other compounds is shown in Fig. 6.19. A schematic structure of the phototransistor is shown in Fig. 6.20. The top source-drain contacts were made using LiF/Al. The devices are dominantly n-type transistors. If PVA is used as gate insulator, gate voltage induced saturation occurred upon illumination. The drain source current is increased by more than two orders of magnitudes upon illumination. The increase occurs due to generation of a large number of free carriers due to photoinduced charge transfer at the conjugated polymer/fullerene bulk heterojunction upon illumination. The dark transfer characteristics change after illumination. The initial dark characteristics can be restored either by applying a large negative gate bias or by annealing. The observed $I–V$ curves of the phototransistor are shown in Fig. 6.21.

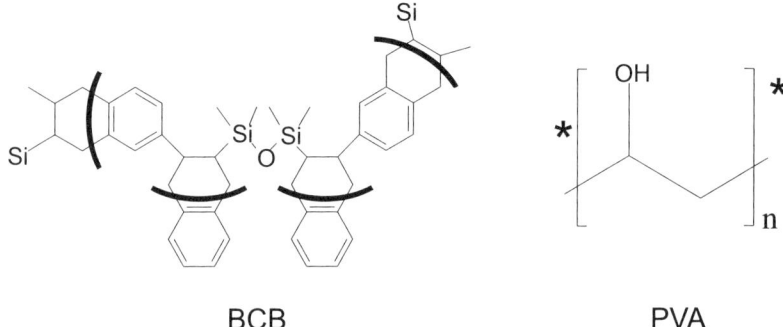

FIG. 6.19. Molecular structure of BCB and PVA; [152].

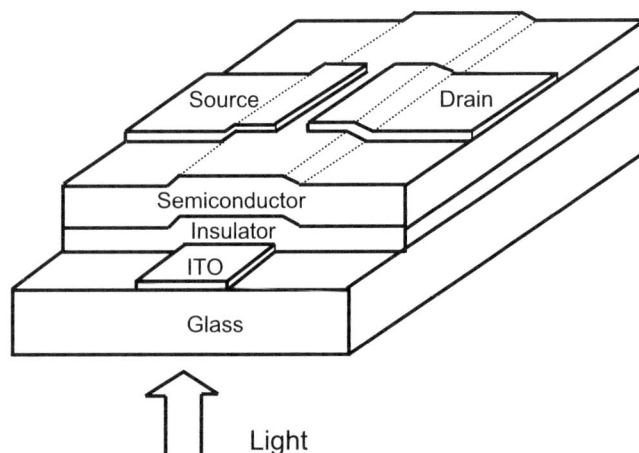

FIG. 6.20. A schematic structure of the phototransistor [152].

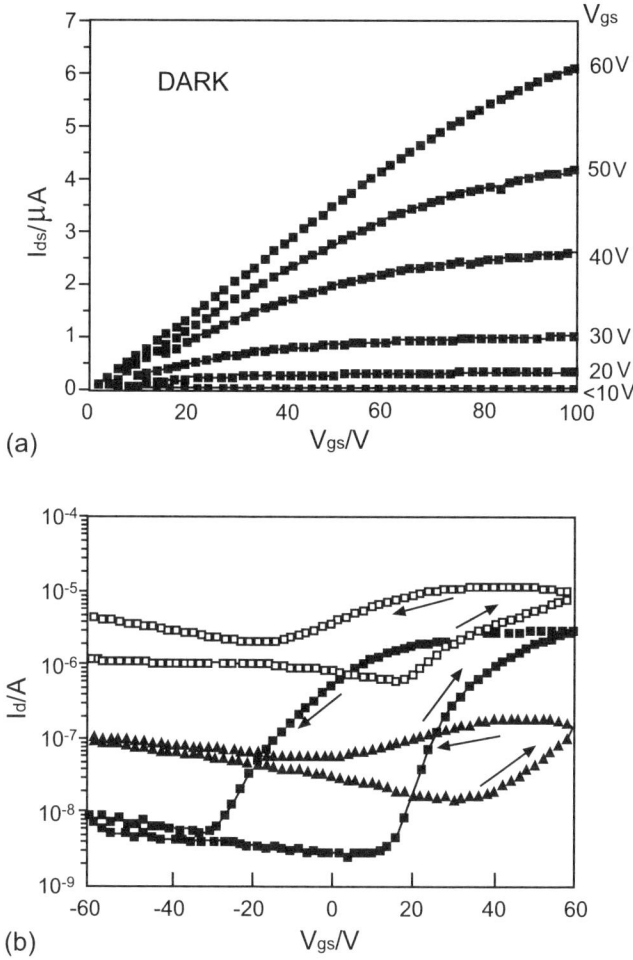

FIG. 6.21. (a) Output characteristics of the MDMO-PPV:PCBM (1:4) based photOFET fabricated on top of a PVA gate-insulator with LiF/Al as top source and drain electrodes in the dark, (b) transfer characteristics of the device in the dark (filled square symbol curves), under AM1.5 (1 mW/cm^2) illumination (open square symbol curves) and in the dark after illumination (filled triangular symbol curves) measured at $V_{ds} = +80$ V. The arrows show the sweep directions [152].

6.9. Organic Dielectrics

Generally pentacene devices are fabricated on silicon wafers with a layer of SiO$_2$ serving as the gate dielectric. Performance of the transistor is improved by the use of a self-assembled monolayer of octadecyltrichlorosilane or similar material [162]. Recently, field effect mobilities up to 3 cm^2/V s in pentacene films deposited on the styrenic polymer poly(4-hydroxystyrene) (PHS) have been reported. The PHS was used both as the substrate and as the gate dielectric. The growth of the pentacene film was done by vapor deposition on PHS. Subsequently, it has been suggested that thin coatings of poly (methylstyrene) on SiO_2 can produce mobilities up to 5 $cm^2/V\,s$ in pentacene devices. In a thin film transistor the current is confined to a very thin region at the interface between the semiconductor and the dielectric. Therefore morphology and chemistry of the dielectric are critical for good device performance. It is known that in amorphous semiconductors, using a low permittivity (nonpolar) fluoropolymer results in substantial improvements in mobility. Nunes et al. [162] investigated if there are materials within the broad range of styrenic polymers which offer performance advantage similar to PHS, and have improved chemical, structural, and electrical properties.

Nunes et al. wanted to find out whether transistors made with PHS and other organic dielectrics could show the excellent performance already reported for the pentacene transistors. This will help to determine whether other polymers with improved characteristics can be found. They investigated the effect of PHS on important device characteristics, threshold voltage and subthreshold slope. The dielectrics they investigated and their dielectric properties are shown in Fig. 6.22. Measured mobilities in transistors fabricated

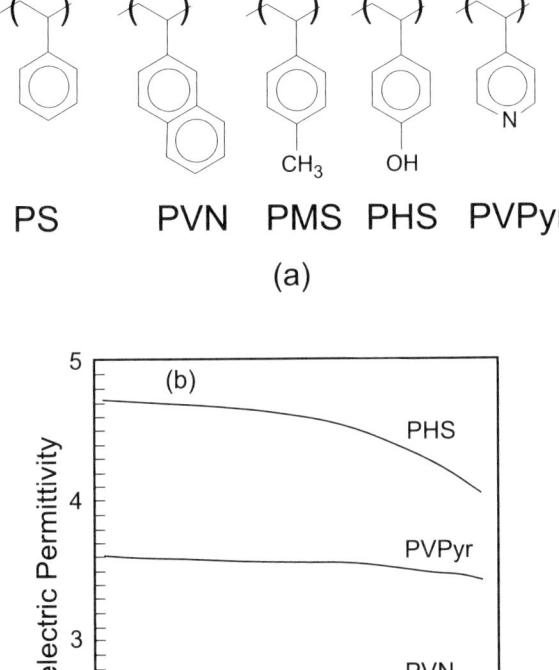

FIG. 6.22. (a) Chemical structure of the polymer dielectrics used by Nunes in his experiments. (b) Measured dielectric permittivities of the five styrenic polymers from 40 Hz to 1 MHz [162].

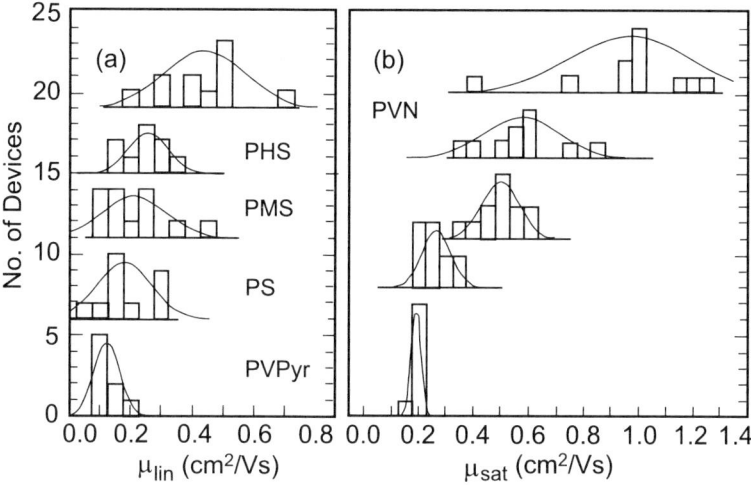

FIG. 6.23. Histograms of the measured (a) linear and (b) saturated regime mobilities for transistors fabricated with styrenic polymer dielectrics. The Gaussian curves are meant only as suggestive of the parent distribution for each measurement set. The histograms are offset vertically for clarity. The order of vertical arrangement is the same in both the left- and right-hand sides of the diagram [162].

with these dielectrics are shown in Fig. 6.23. Transistors made with PHS showed high mobility. However there are other polymers which offer better device characteristics, mainly because of their nonpolar nature. In transistors fabricated with the use of PMS as dielectric showed mobility comparable to that of the PHS devices. At the same time there was significant improvement in the other characteristics. The mobility for PVPyr devices was the lowest observed. The average of the V_t measurements for PVPyr was practically zero. This indicates relatively fewer traps at the interface. Further work is necessary to determine whether a surface treatment and/or different set of deposition conditions could be found which would improve all the characteristics.

The AFM pictures of the pentacene films on styrenic polymers in the channel region of transistors are shown in Fig. 6.24. The mobility observed on PHS is lower than the best values reported in the literature. It may be possible to obtain improvements in the mobility values without degrading other characteristics. The average mobility for these devices was almost twice that observed for the PHS devices, implying that pentacene transistors with a field effect mobility in the range of 6–10 cm^2/V s should be achievable, which would expand very significantly the range of applications for organic TFTs. In order for these devices to be practical, however, the source of the scatter would have to be significantly reduced. In fact, though transistors fabricated using PS, PMS, and PVPyr showed a much narrower distribution in device properties than those fabricated with PVN and PHS, the variation in all cases is too large for a manufacturing process. To make organic, flexible electronics a practical reality, the sources of this variation must be determined and controlled.

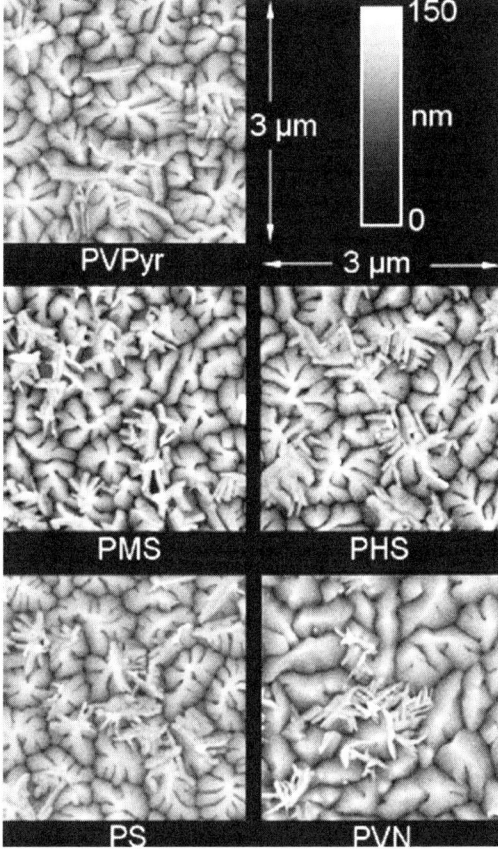

FIG. 6.24. AFM images of pentacene growth on styrenic polymers in the channel region of transistors. Most of the substrates show crystallites of the usual dendritic form, with the exception of PVN, where the dendritic growth appears to be somewhat suppressed. The needle-like growths visible in the image are most likely pentacene dihydride [162].

Bibliography

[1] H. Ehrenreich, F. Spaepen (Eds.), *Solid State Physics, Advances in Research and Applications*, Academic Press, San Diego, 1994.
[2] R.B. Kanner, A.G. MacDiarmid, Plastics that conduct electricity, *Scientific American* (February 1988) 60–65.
[3] P. Yam, Plastics get wired, *Scientific American* (July 1995) 75–79.
[4] M. Kenward, Polymer based displays, *Compound Semiconductors* (January/February 1997) 30–33.
[5] G. Wicks, Organic lasers, *Compound Semiconductors* (January/February 1997) 42.
[6] A.J. Heeger, S. Kivelson, J.R. Schrieffer, W.-P. Su, Solitons in conducting polymers, *Rev. Mod. Phys.* **60** (1988) 781–850.
[7] T.A. Skotheim (Ed.), *Handbook of Conducting Polymers, Vols. 1 and 2*, Marcel Dekker Inc., 1986.
[8] A.J. Heeger, Polyacetylene: New concepts and phenomena, 1986, in Ref. [7, pp. 729–756].
[9] R. Menon, Charge transport in conducting polymers, in: Hari Singh Nalwa (Ed.), *Handbook of Organic Conductive Molecules and Polymers*, Vol. 4, John Wiley & Sons, 1997, pp. 47–145.
[10] S. Roth, W. Graupner, *Synth. Met.* **55–57** (1993) 3623.
[11] S.R. Forrest, Ultra thin organic molecular beam deposition and related techniques, *Chem. Rev.* **97** (1997) 1793–1896.
[12] A.J. Epstein, Electrically conducting polymers: Science and technology, *MRS Bulletin* (June 1997) 16–23.
[13] R.S. Kohlman, et al., Limits for metallic conductivity in conducting polymers, *Phys. Rev. Lett.* **78** (1997) 3915–3918.
[14] E.M. Conwell, H.A. Mizes, Conjugated polymer semiconductors: An introduction, in: P.T. Landsberg (Vol. Ed.), T.S. Moss (Ed.), *Handbook of Semiconductors*, Vol. 1, Elsevier Science Publishers B.V., 1992, pp. 584–625.
[15] R.N. Marks, J.J.M. Halls, D.D.C. Bradley, R.H. Friend, A.B. Holmes, The photovoltaic response in poly(p-phenylene vinylene) thin-film devices, *J. Phys. Condens. Matter* **6** (1994) 1379–1394.
[16] H. Koezuka, *Synth. Met.* **18** (1987) 699.
[17] E.M. Conwell, Transport in conducting polymers, in: Hari Singh Nalwa (Ed.), *Handbook of Organic Conductive Molecules and Polymers*, Vol. 4, John Wiley & Sons, 1997, pp. 1–46.

[18] A. Feldblum, J.H. Kaufman, S. Etemad, A.J. Heeger, T.-C. Chung, A.G. MacDiarmid, *Phys. Rev. B* **26** (1982) 815.
[19] M. Takayama, Y.R. Lin-Liu, K. Maki, *Phys. Rev. B* **21** (1980) 2388.
[20] N. Suzuki, M. Ozaki, S. Etemad, A.J. Heeger, A.G. MacDiarmid, *Phys. Rev. Lett.* **45** (1980) 1209; Erratum, *Phys. Rev. Lett.* **45** (1980) 1453.
[21] G.B. Blanchet, C.R. Fincher, T.-C. Chung, A.J. Heeger, *Phys. Rev. Lett.* **50** (1983) 1938.
[22] S. Etemad, M. Mitani, M. Ozaki, T.-C. Chung, A.J. Heeger, A.G. MacDiarmid, *Solid State Commun.* **40** (1981) 1656–1666.
[23] S. Jeyadev, E.M. Conwell, *Phys. Rev. Lett.* **58** (1987) 258; *Phys. Rev. B* **36** (1987) 3284.
[24] M. Nechtschein, F. Devreux, F. Genoud, M. Guglielmi, K. Holczer, *Phys. Rev. B* **27** (1983) 61.
[25] S. Jeyadev, E.M. Conwell, *Mol. Cryst. & Liq. Cryst.* **160** (1988) 443.
[26] K. Mizoguchi, H. Shirakawa, *Abstracts Int. Conf. on Science and Technology of Synthetic Metals*, Kyoto, 1986, p. 55.
[27] J. Tsukamoto, A. Takahashi, K. Kawasaki, *Jpn. J. Appl. Phys.* **29** (1990) 125.
[28] K. Ehinger, S. Roth, *Phil. Mag. B* **53** (1986) 301.
[29] R. Kersting, et al., Ultrafast field-induced dissociation of excitons in conjugated polymers, *Phys. Rev. Lett.* **73** (1994) 1440–1443;
V.I. Arkhipov, et al., Field induced exciton breaking in conjugated polymers, *Phys. Rev. B* **52** (1995) 4932–4940.
[30] V. Kumar, S.C. Jain, T. Aernouts, W. Geens, R. Mertens, unpublished.
[31] S.C. Binari, H.C. Dietrich, in: S.J. Pearton (Ed.), *GaN and Related Materials*, Gordon and Breach Science Publishers, 1997, pp. 509–534.
[32] L. Pfeiffer, K.W. West, H.L. Stormer, K.W. Baldwin, *Appl. Phys. Lett.* **55** (1989) 1888.
[33] M.A. Khan, Q. Chen, J.W. Yang, C.J. Sun, *Inst. Phys. Conf. Ser.* **142** (1995) 985–990.
[34] G. Yu, A.J. Heeger, Charge separation and photovoltaic conversion in polymer composites with internal donor/acceptor heterojunctions, *J. Appl. Phys.* **78** (1995) 4510–4515.
[35] K. Yoshino, K. Tada, A. Fujii, E.M. Conwell, A. Zakhidov, Novel photovoltaic devices based on donor–acceptor molecular and conducting polymer systems, *IEEE Trans. Electron Devices* **44** (1997) 1315–1324.
[36] N.F. Mott, R.W. Gurney, *Electronic Processes in Ionic Crystals*, Dover Publications, Inc., New York, 1940. Reprinted in 1964.
[37] M.A. Lampert, P. Mark, *Current Injection in Solids*, Academic Press, New York, 1970.
[38] K.C. Kao, W. Hwang, *Electrical Transport in Solids*, Pergamon Press, Oxford, 1981.
[39] V. Kumar, S.C. Jain, A.K. Kapoor, W. Geens, T. Aernouts, J. Poortmans, R. Mertens, Carrier transport in conducting polymers with field dependent trap occupancy, *J. Appl. Phys.* **92** (2002) 7325–7329.
[40] S.C. Jain, W. Geens, A. Mehra, V. Kumar, T. Aernouts, J. Poortmans, R. Mertens, Injection- and space charge limited-currents in doped conducting organic materials, *J. Appl. Phys.* **89** (2001) 3804–3810.

[41] S.C. Jain, A.K. Kapoor, W. Geens, J. Poortmans, R. Mertens, M. Willander, Trap filled limit of conducting organic materials, *J. Appl. Phys.* **92** (2002) 3752–3754.

[42] A.K. Kapoor, S.C. Jain, J. Poortmans, V. Kumar, R. Mertens, Temperature dependence of carrier transport in conducting polymers: Similarity to amorphous inorganic semiconductors, *J. Appl. Phys.* **92** (2002) 3835.

[43] S.C. Jain, T. Aernouts, A.K. Kapoor, V. Kumar, W. Geens, J. Poortmans, R. Mertens, I-V Characteristics of dark and illuminated PPV-PCBM blends solar cells, *Synth. Met.* **148** (2005) 245–250.

[44] P. Kumar, S.C. Jain, A. Misra, M.N. Kamalasanan, V. Kumar, Characteristics of a conducting organic diode with finite (non-zero) Schottky barrier, *J. Appl. Phys.* **100** (2006) 114506.

[45] S.M. Sze, *Physics of Semiconductor Devices*, John Wiley and Sons, New York, 1981.

[46] A.J. Campbell, D.D.C. Bradley, D.G. Lidzey, Space-charge limited conduction with traps in poly(phenylene vinylene) light emitting diodes, *J. Appl. Phys.* **82** (1997) 6326–6342;
A.J. Campbell, M.S. Weaver, D.G. Lidzey, D.D.C. Bradley, Bulk limited conduction in electroluminescent polymer devices, *J. Appl. Phys.* **84** (1998) 6737–6746.

[47] S. Berleb, A.G. Mückl, W. Brütting, M. Schwoerer, *Synth. Met.* **111–112** (2000) 341.

[48] J.M. Lupton, I.D.W. Samuel, *Synth. Met.* **111–112** (2000) 381.

[49] V. Kumar, S.C. Jain, A.K. Kapoor, J. Poortmans, R. Mertens, Trap density in conducting organic semiconductors determined from temperature dependence of $J-V$ characteristics, *J. Appl. Phys.* **94** (2003) 1283–1285.

[50] I.D. Parker, Carrier tunnelling and device characteristics in polymer light-emitting diodes, *J. Appl. Phys.* **75** (1994) 1951–1960.

[51] P.W.M. Blom, M.J.M. de Long, M.G. van Munster, *Phys. Rev. B* **55** (1997) R656.

[52] A.J. Campbell, D.D.C. Bradley, T. Virgili, D.G. Lidzey, H. Antoniadis, *Appl. Phys. Lett.* **79** (2001) 3872.

[53] P. Kumar, A. Misra, M.N. Kamalasanan, S.C. Jain, R. Srivastava, V. Kumar, Charge transport through conducting organic poly(2-methoxy-5-(2-ethylhexyloxy)-1,4-phenylene vinylene), *J. Phys. D: Appl. Phys.* **40** (2007) 561.

[54] P. Kumar, A. Misra, M.N. Kamalasanan, S.C. Jain, R. Srivastava, V. Kumar, *Jpn. J. Appl. Phys. 10A* **45** (2006) 7621.

[55] P.E. Burrows, Z. Shen, V. Bulović, D.M. McCarty, S.R. Forrest, J.A. Cronin, M.E. Thompson, *J. Appl. Phys.* **79** (1996) 7991.

[56] S.C. Jain, W. Geens, V. Kumar, A. Kapoor, A. Mehra, T. Aernouts, J. Poortmans, R. Mertens, M. Willander, A unified model for the space charge limited currents in organic materials combining field dependent mobility and Poole–Frenkel detrapping, in: MRS Spring 2001, San Francisco, in: *Mat. Res. Soc. Symp. Proc.*, Vol. 665, 2001, pp. C8.12.1–C8.12.7.

[57] P.W.M. Blom, M.J.M. de Jong, *IEEE J. Selected Topics in Quantum Electronics* **4** (1996) 105.

[58] P.C. Arnett, N. Klein, *J. Appl. Phys.* **46** (1975) 1399.

[59] B.K. Crone, I.H. Campbell, P.S. Davids, D.L. Smith, *J. Appl. Phys.* **86** (1999) 5767–5774.

[60] C. Melzer, E.J. Koop, V.D. Mihailetchi, P.W.M. Blom, Hole transport in poly(phenylene vinylene)/methanofullerene bulk-heterojunction solar cells, *Adv. Funct. Mater.* **14** (2004) 865–870.
[61] G.H. Hewig, W.H. Bloss, Technology of thin films solar cells, *Thin Solid Films* **45** (1977) 1–7.
[62] G. Agostinelli, Photocurrent Analysis of CdTe Solar Cells, Ph.D. Thesis, Universiteit Gent, 2002. See also: G. Agostinelli, et al., Light dependent current transport mechanisms in chalcogenide solar cells, in: *3rd World Photovoltaic Energy Conference*, Osaka, May 11–18, 2003.
[63] V.D. Mihailetchi, et al., *Adv. Funct. Mater.* **13** (2003) 43.
[64] M. Kermerink, et al., *Nano Lett.* **3** (2003) 1191.
[65] J.K.J. van Duren, V.D. Mihailetchi, P.W.M. Blom, T. van Woudenbergh, J.C. Hummelen, M.T. Rispens, R.A.J. Janssen, M.M. Wienk, Injection-limited electron current in a methanofullerene, *J. Appl. Phys.* **94** (2003) 4477–4479.
[66] C.W. Tang, S.A. VanSlyke, *Appl. Phys. Lett.* **51** (1987) 913.
[67] J.H. Burroughes, D.D.C. Bradley, A.R. Brown, R.N. Marks, K. Mackay, R.H. Friend, P.L. Burn, A.B. Holmes, Light emitting diodes based on conjugated polymers, *Nature* **347** (1990) 539–541.
[68] F. Li, H. Tang, J. Shinar, O. Resto, S.Z. Weisz, *Appl. Phys. Lett.* **70** (1997) 2741.
[69] M. Kenward, *Compound Semiconductors* (1996).
[70] R.H. Friend, R.W. Gymer, A.B. Holmes, J.H. Burroughes, R.N. Marks, C. Taliani, D.D.C. Bradley, D.A.D. Santos, J.L. Bredas, M. Logdlund, W.R. Salaneck, *Nature* **397** (1999) 121.
[71] M. Berggren, A. Dodabalapur, R.E. Slusher, Z. Bao, Light amplification in organic thin films using cascade energy transfer, *Nature* **389** (1997) 466–469.
[72] P.E. Burrows, G. Gu, S.R. Forrest, *SPIE* **3363** (1998) 269.
[73] U. Lemmer, *Polym. Adv. Technol.* **9** (1998) 476.
[74] X.Y. Jiang, Z.L. Zhang, B.X. Zhang, W.Q. Zhu, S.H. Xu, *Synth. Met.* **129** (2002) 9.
[75] Y. Wang, F. Teng, C. Ma, Z. Xu, Y. Hou, S. Yang, Y. Wang, X. Xu, *Display* **25** (2004) 237.
[76] C. Legnani, R. Reyes, M. Cremona, I.A. Bagatin, H.E. Toma, *Appl. Phys. Lett.* **85** (2004) 10.
[77] G. Mauthner, M. Collon, E.J.W. List, F.P. Wenzl, M. Bouguettaya, J.R. Reynolds, *J. Appl. Phys.* **97** (2005) 63508.
[78] P.E. Rurrows, G. Gu, V. Buloviæ, *IEEE Trans. Electron Devices* **44** (1997) 1188.
[79] C.C. Wu, Y.T. Lin, H.H. Chiang, T.Y. Cho, C.W. Chen, K.T. Wong, Y.L. Liao, G.H. Lee, S.M. Peng, *Appl. Phys. Lett.* **81** (2002) 577.
[80] W.B. Im, K. Hwang, G. Lee, K. Han, Y. Kim, *Appl. Phys. Lett.* **79** (2001) 1387.
[81] M. Mazzeo, D. Pisignano, L. Favaretto, G. Sotgiu, G. Barbarella, R. Cingolani, G. Gigli, *Synth. Met.* **139** (2003) 675.
[82] A. Islam, P. Murugan, K.C. Hwang, C.H. Cheng, *Synth. Met.* **139** (2003) 347.
[83] Y. Kan, L. Wang, Y. Gao, L. Duan, G. Wu, Y. Qiu, *Synth. Met.* **141** (2004) 245.
[84] E.C. Chang, S.A. Chen, *J. Appl. Phys.* **85** (1999) 2057.
[85] A. Costela, et al., *Appl. Phys. B* **67** (1998) 167.
[86] L.M. Leung, W.Y. Lo, S.K. So, K.M. Lee, W.K. Choi, *J. Am. Chem. Soc.* **122** (2000) 5640.

[87] Y.Z. Wu, X.Y. Zheng, W.Q. Zhu, R.G. Sun, X.Y. Jiang, Z.L. Zhang, S.H. Xu, *Appl. Phys. Lett.* **83** (2003) 5077.
[88] Y. Kijima, N. Asai, S. Tamura, *Jpn. J. Appl. Phys.* **138** (9A) (1999) 5274.
[89] S. Tokito, T. Tsuzuki, F. Sato, T. Iijima, *Curr. Appl. Phys.* **5** (2005) 331.
[90] R.I. Laskar, S.F. Hsu, T.M. Chen, *Polyhedron* **24** (2005) 189.
[91] T. Li, T. Yamamoto, H.L. Lan, J. Kido, *Polym. Adv. Technol.* **15** (2004) 266.
[92] Y. Yang, Polymer electroluminescent devices, 1997, in Ref. [125, pp. 31–38].
[93] A. Debray, et al., *Components* (1998) 20–21.
[94] S. Nakamura, T. Mukai, M. Senoh, *Appl. Phys. Lett.* **64** (1994) 1687–1689.
[95] S.P. Singh, Y.N. Mohapatra, M. Qureshi, S. Manoharan, *Appl. Phys. Lett.* **86** (2005) 113505.
[96] A.R. Duggal, J.J. Shiang, C.M. Heller, D.F. Foust, *Appl. Phys. Lett.* **80** (2002) 3470.
[97] A. Dodabalapur, L.J. Rothberg, T. Miller, *Appl. Phys. Lett.* **65** (1994) 2308.
[98] M. Mazzeo, V. Vitale, F.D. Sala, M. Anni, G. Barbarella, L. Favaretto, G. Sotgiu, R. Cingolani, G. Gigli, *Adv. Mat.* **17** (2005) 34.
[99] D.B. Eason, Z. Yu, W.C. Hughes, W.H. Roland, C. Boney, J.W. Cook Jr., J.F. Schetzina, *Appl. Phys. Lett.* **66** (1995) 115.
[100] T. Strite, W. Riess, *Compound Semiconductors* (November/December 1997) 34–38.
[101] A. Dodabalapur, E.A. Chandros, M. Berggren, R.E. Slusher, Organic solid state lasers: Past and future, *Science* **277** (1997) 1787–1788.
[102] Z.Y. Xie, L.S. Hung, *Appl. Phys. Lett.* **84** (2004) 1207.
[103] S.V. Frolov, W. Gellerman, M. Ozaki, K. Yoshino, Z.V. Vardeny, *Phys. Rev. Lett.* **78** (1997) 729.
[104] M.D. McGehee, R. Gupta, S. Veenstra, E.K. Miller, M.A. Diaz-Garcia, A.J. Heeger, *Phys. Rev. B* **58** (1998) 7035.
[105] Y.G. Kozlov, G. Parthasarthy, P.E. Burrows, S.R. Forest, Y. You, M.E. Thompson, Optically pumped blue organic semiconductor lasers, *Appl. Phys. Lett.* **72** (1998) 144–146.
[106] T.R. Hebner, C.C. Wu, D. Marcy, M.H. Lu, J.C. Sturm, Ink-jet printing of doped polymers for organic light emitting devices, *Appl. Phys. Lett.* **72** (1998) 519–521.
[107] S.C. Jain, *Germanium–Silicon Strained Layers and Heterostructures*, P.W. Hawkes (Ed.), *Advances in Electronics and Electron Physics Series*, Supplement 24, Academic Press, Boston, 1994.
[108] J. Liu, Y. Shi, L. Ma, Y. Yang, Device performance and polymer morphology in polymer light emitting diodes: The control of device electrical properties and metal/polymer contact, *J. Appl. Phys.* **88** (2000) 605–609.
[109] A.J.M. Berntsen, P. Van de Weirger, Y. Croonen, C.T.H.F. Liedenbaum, J.J.M. Vleggaar, Stability of polymer light-emitting diodes, *Philips J. Res.* **51** (1998) 511–525.
[110] M. Schaer, F. Nüesch, D. Berner, W. Leo, L. Zuppirol, Water vapor and oxygen degradation mechanisms in organic light emitting diodes, *Adv. Funct. Mater.* **11** (2001) 116–121.
[111] R. Czerw, D.L. Carroll, H.S. Woo, Y.B. Kim, J.W. Park, Nanoscale observation of failures in organic light-emitting diodes, *J. Appl. Phys.* **96** (2004) 641–644.
[112] S.M. Jeong, et al., Charge injection and transport model in organic ligt emitting diodes, *Solid-State Electronics* **49** (2005) 205–212.

[113] A. Misra, P. Kumar, M.N. Kamalasanan, S.C. Jain, Synthesis and characterization of some 5-coordinated aluminum-8-hydroxyquinoline derivatives for OLED applications, to be published.

[114] J. Kanicki, Polymeric semiconductor contacts and photovoltaic applications, 1986, in Ref. [7, pp. 543–660].

[115] L.S. Hunga, C.H. Chen, Recent progress of molecular organic electroluminescent materials and devices, *Materials Science and Engineering R* **39** (2002) 143–222.

[116] N.S. Sariciftci, S. Smilowitz, A.J. Heeger, F. Wudl, Photoinduced electron transfer from a conducting polymer to buckminsterfullerene, *Science* **258** (1992) 1474–1476.

[117] S. Morita, A.A. Zakhidov, K. Yoshino, *Solid State Commun.* **82** (1992) 249–252.

[118] G. Yu, K. Pakbaz, A.J. Heeger, Semiconducting polymer diodes: Large size, low cost photodetectors with excellent visible-ultraviolet sensitivity, *Appl. Phys. Lett.* **64** (1994) 3422–3424.

[119] G. Yu, J. Gao, J.C. Hummelen, F. Wudl, A.J. Heeger, Polymer photovoltaic cells: Enhanced efficiencies via a network of internal donor–acceptor heterojunctions, *Science* **270** (1995) 1789–1791.

[120] X. Wei, M. Raikh, Z.V. Vardeny, Y. Yang, D. Moses, *Phys. Rev. B* **49** (1994) 17480.

[121] J. Gao, F. Hide, H. Wang, Efficient photodetectors and photovoltaic cells from composites of fullerenes and conjugated polymers: photoinduced electron transfer, *Synth. Met.* **84** (1997) 979–980.

[122] J.J.M. Halls, C.A. Walsh, N.C. Greenham, E.A. Marseglla, R.H. Friend, S.C. Moratti, A.B. Holmes, Efficient photodiodes from interpenetrating polymer networks, *Nature* **376** (1995) 498–500.

[123] B.A. Gregg, Photovoltaic properties of a molecular semiconductor modulated by an exciton-dissociating film, *Appl. Phys. Lett.* **67** (1995) 1271–1273.

[124] C.J. Brabec, S.E. Shaheen, C. Winder, N.S Sariciftci, P. Denk, Effect of LiF metal electrodes on the performance of plastic solar cells, *Appl. Phys. Lett.* **80** (2002) 1288.

[125] Special volume of MRS Bulletin on conducting polymers, *MRS Bulletin* (June 1997).

[126] H. Spanggaard, F.C. Krebs, A brief history of the development of organic and polymeric photovoltaics, *Solar Energy Materials and Solar Cells* **83** (2004) 125–146.

[127] S.C. Jain, W. Geens, J. Poortmans, A. Mehra, J. Nijs, R. Mertens, Exciton dissociation and mobility in conducting polymers and oligomers, in: San Francisco, MRS Spring Meeting 1999, in: *Mat. Res. Soc. Symp. Proc.*, Vol. 558, 2000, pp. 485–490.

[128] A.J. Breeze, A. Salomon, D.S. Ginley, B.A. Gregg, H. Tillmann, H.-H. Horhold, Polymer-perylene diimide heterojunction solar cells, *Appl. Phys. Lett.* **81** (2002) 3085–3087.

[129] F. Padinger, R.S. Rittberger, S. Sariciftci, Effect of post production heat treatment on plastic solar cells, *Adv. Funct. Mater.* **13** (2003) 85–88.

[130] Siemens researchers achieve a breakthrough in increasing the efficiency of organic solar cells, News release for the Trade Press by Siemens, Munich, 7 January 2004. C E Printed plastic solar cells.

[131] E. Kymakis, G.A.J. Amaratunga, Single-wall carbon nanotube/conjugated polymer photovoltaic devices, *Appl. Phys. Lett.* **80** (2002) 112–124.

[132] E. Kymakis, I. Alexandu, G.A.J. Amaratunga, Single-walled carbon nanotube-polymer composites: Electrical, optical and structural investigation, *Synth. Met.* **127** (2002) 59–62.

[133] C.J. Brabec, A. Cravino, D. Meissner, et al., Origin of the open circuit voltage of plastic solar cells, *Adv. Funct. Mater.* **11** (2001) 374–380.

[134] V.D. Mihailetchi, P.W.M. Blom, J.C. Hummelen, M.T. Rispens, Cathode dependence of the open-circuit voltage of polymer:fullerene bulk heterojunction solar cells, *J. Appl. Phys.* **94** (2003) 6849–6854.

[135] C.M. Ramsdale, J.A. Barker, A.C. Arias, J.D. MacKenzie, R.H. Friend, N.C. Greenham, The origin of the open-circuit voltage in polyfluorene-based photovoltaic devices, *J. Appl. Phys.* **92** (2002) 4266–4270.

[136] J.K.J. vanDuren, X. Yang, J. Loos, C.W.T. Bulle-Lieuwma, A.B. Sieval, J.C. Hummelen, R.A.J. Janssen, *Adv. Funct. Mater.* **14** (2004) 425.

[137] H.P. Maruska, T.D. Mustakas, Influence of the wavelength of incident light on shunt conductance and fill factor in amorphous silicon solar cells, *IEEE Trans. Electron Devices* **31** (1984) 551–558.

[138] F.A. Lindholm, J.G. Fossum, E.L. Burgess, Basic corrections to predictions of solar cell performance required by non-linearities, in: *Proc. 12th Photovoltaic Specialists Conf.*, 1976, pp. 33–39.

[139] A. Rothwarf, The superposition principle for currents in solar cells, in: *Proc. 13th Photovoltaic Specialists Conf.*, 1978, pp. 1312–1317.

[140] S.C. Jain, E.L. Heasell, D.J. Roulston, Recent advances in the physics of silicon P-N junction solar cells, in: T.S. Moss, et al. (Eds.), *Progress in Quantum Electronics*, Pergamon Press, Oxford, 1987, pp. 105–204.

[141] P. Schilinsky, C. Waldauf, J. Hauch, C.J. Brabec, Simulation of light intensity dependent current characteristics of polymer solar cells, *J. Appl. Phys.* **95** (2004) 2816.

[142] M.S. Yoo, B. Domercq, B. Kippelen, Efficient thin-film organic solar cells based on pentacene/C60 heterojunctions, *Appl. Phys. Lett.* **85** (2004) 5427–5429.

[143] M. Pfeiffer, A. Beyer, B. Plonnigs, A. Nollau, T. Fritz, K. Leo, D. Schlettwein, S. Hiller, D. Wohrle, Controlled p-doping of pigment layers by Cosublimation: Basic mechanisms and implications for their use in organic photovoltaic cells, *Solar Energy Materials and Solar Cells* **63** (2000) 83–99.

[144] M. Guldi, et al., Molecular Engineering of C60 based oligomer ensembles: Modulating the competition between photoinduced energy and electron transfer processes, *J. Org. Chem.* **67** (2002) 1141–1152.

[145] Jiangeng Xue, S. Uchida, B.P. Rand, Stephen R. Forrest, Asymmetric tandem organic photovoltaic cells with hybrid planar-mixed molecular heterojunctions, *Appl. Phys. Lett.* **85** (2004) 5757–5759.

[146] S.S. Hegedus, N. Salzman, E. Fagen, The relation of dark and illuminated diode parameters to the open-circuit voltage of amorphous silicon p-i-n solar cells, *J. Appl. Phys.* **63** (1988) 5126–5130.

[147] M.K. Han, P. Sung, W.A. Anderson, Determination of built-in-potential in N-I-P a-Si:H solar cells, *IEEE Electron Device Letters* **3** (1982) 121–124.

[148] S. Dhar, V. Balakrishnan, V. Kumar, S. Ghosh, Determination of energetic distribution of interface states between gate metal and semiconductor in sub-micron deices

from current-voltage characteristics, *IEEE Trans. Electron Devices* **47** (2000) 282–287.
[149] Z. Chen, L.C. Burton, Electrical properties and their depth variation in poly-Si under AM1 illumination, *Phys. Stat. Sol.* **122** (1990) 361–370.
[150] K.M. Koliwad, T. Daud, Grain size dependence of silicon solar cell parameters, in: *15th IEEE Photovoltaic Specialists Conference*, 1980, pp. 1204–1208.
[151] A.K. Ghosh, C. Fishman, T. Feng, Theory of the electrical and photovoltaic properties of polycrystalline silicon, *J. Appl. Phys.* **51** (1980) 446–454.
[152] N. Marjanovic, et al., Photoresponse of organic field-effect transistors based on conjugated polymer/fullerene blends, *Organic Electronics* **7** (2006) 188–194.
[153] W.Y. Chou, C.W. Kuo, H.L. Cheng, Y.S. Mai, F.C. Tang, S.T. Lin, C.Y. Yeh, J.B. Horng, Epitaxial pentacene films grown on the surface of ion-beam-processed gate dielectric layer, in: C.T. Chia, C.C. Liao, D.Y. Shu (Eds.), *J. Appl. Phys.* **99** (2006) 114511.
[154] F. De Angelis, L. Mariucci, S. Cipolloni, G. Fortunato, Analysis of electrical characteristics of high performance pentacene thin-film transistors with PMMA buffer layer, *Journal of Non-Crystalline Solids* **352** (2006) 1765–1768.
[155] A. Dodabalapur, L. Tosi, H.E. Katz, Organic transistors: Two dimensional transport and improved electrical characteristics, *Science* **268** (1995) 270–271.
[156] G. Horowitz, *Adv. Mat.* **2** (1990) 287.
[157] G. Horowitz, P. Delannoy, An analytical model for organic-based thin-film transistors, *J. Appl. Phys.* **70** (1991) 469–475.
[158] Rashmi, V.R. Balakrishnan, A.K. Kapoor, V. Kumar, S.C. Jain, R. Mertens, S. Annapoorni, Effect of field dependent trap occupancy on organic thin film transistor characteristics, *J. Appl. Phys.* **94** (2003) 5302–5306.
[159] F. Garnier, R. Hajlaoui, A. Yassar, P. Srivastava, All-polymer field-effect transistor realized by printing technique, *Science* **265** (1994) 1684–1686.
[160] H. Klauk, Y.Y. Lin, D.J. Gundlach, T.N. Jackson, *International Electron Devices Meeting Technical Digest*, 1997, p. 539.
[161] G. Horowitz, Tunneling current in polycrystalline organic thin film transistor, *Adv. Funct. Mater.* **13** (2003) 53–60.
[162] G. Nunes Jr., S.G. Zane, J.S. Methb, Styrenic polymers as gate dielectrics for pentacene field-effect transistors, *J. Appl. Phys.* **98** (2005) 104503.
[163] D.J. Gundlach, L. Zhou, J.A. Nichols, T.N. Jackson, P.V. Necliudovc, M.S. Shur, An experimental study of contact effects in organic thin film transistors, *J. Appl. Phys.* **100** (2006) 24509.
[164] Mohammad Mottaghi, Gilles Horowitz, Field-induced mobility degradation in pentacene thin-film transistors, *Organic Electronics* (2006), article in press, available on-line at Science Direct.
[165] D. Guo, S. Ikeda, K. Saiki, H. Miyazoe, K. Terashima, Effect of annealing on the mobility and morphology of thermally activated pentacene thin film transistors, *J. Appl. Phys.* **99** (2006) 094502.
[166] J. Liu, Y. Shi, Y. Yang, *Adv. Funct. Mater.* **11** (2001) 420.
[167] R. Kersting, et al., Ultrafast field-induced dissociation of excitons in conjugated polymers, *Phys. Rev. Lett.* **73** (1994) 1440–1443;
V.I. Arkhipov, et al., Field induced exciton breaking in conjugated polymers, *Phys. Rev. B* **52** (1995) 4932–4940.

[168] R.S. Deshpande, V. Buloviæ, S.R. Forrest, *Appl. Phys. Lett.* **75** (1999) 888.
[169] A.W. Grice, et al., *Appl. Phys. Lett.* **73** (1998) 629.
[170] S. Nakamura, in: S.J. Pearton (Ed.), *GaN and Related Materials*, Gordon and Breach Science Publishers, 1997, pp. 471–507.
[171] S. Nakamura, *Selected Topics in Quantum Electronics* **3** (1997) 712.
[172] E. Kato, H. Noguchi, M. Nagai, H. Okuyama, S. Kijima, A. Ishibashi, *Electronics Letters* **34** (1998) 282–284.
[173] S. Nakamura, M. Senoh, S. Nagahama, N. Iwasa, T. Matushita, T. Mukai, *MRS Internet J. Nitride Semicond. Res.* **4S1** (1999) G1.1.
[174] S. Stagira, M. Zavelani-Rossi, M. Nisoli, S. DeSilvestri, G. Lanzani, C. Zenz, P. Mataloni, G. Leising, Single-mode picosecond blue laser emission from a solid conjugated polymer, *Appl. Phys. Lett.* **73** (1998).
[175] A. Dodabalapur, M. Berggren, R.E. Slusher, Z. Bao, A. Timko, P. Schiortino, E. Laskowski, H.E. Katz, O. Nalamasu, Resonators and materials for organic lasers based on energy transfer, *IEEE Journal of Selected Topics in Quantum Electronics* **4** (1998) 67–74.
[176] R.L. Gunshor, A.V. Nurmikko (Eds.), *II-VI Blue/Green Light Emitters: Device Physics and Epitaxial Growth*, Academic Press, 1997.
[177] J.I. Pankove, T.D. Moustakas (Eds.), *GaN*, Vol. 1, Academic Press, New York, 1998.
[178] J.C. Scott, J.H. Kaufman, P.J. Brock, R. DiPietro, J. Salem, J.A. Goitia, Degradation and failure of MEH-PPV light-emitting diodes, *J. Appl. Phys.* **79** (1996) 2745–2751.
[179] M. Stolka, *Organic Light Emitting Diodes (OLEDs) for General Illumination*, OIDA Industrial Development Association, USA, Update 2002.
[180] A.R. Tameev, Z. He, G.H.W. Milburn, A.A. Kozlov, A.V. Vannikov, A. Puchala, D. Rasala, Electron drift mobility in polystyrene doped with bispyrazolopyridine derivatives, *Appl. Phys. Lett.* **81** (2002) 969–971.

Index

A

absorption 10, 14, 16, 24, 70, 71, 83, 91, 97, 99
   spectra 77, 96, 99–101
acceptor strength 104–106
acceptors 8, 14, 24, 77, 85, 96, 97, 99, 104–106, 122
activation energy 46, 47, 51, 54
   effective 47, 48
   zero field 54
active devices 4
alignment method 136
alloys 62, 70, 101, 103
aluminum 51, 67, 72, 74, 91, 92, 94, 136, 138
amorphous 5, 32, 110, 116–119, 123–126, 133, 144
amplified spontaneous emission 82
anisotropic 20, 136
annealing 138, 139, 142
anode 51, 62, 70, 86, 90, 106
antibonding 9
antisoliton 15, 24, 26
antistatic 4
applications 2–5, 17, 23, 75, 86, 116, 123, 131, 145
applied
   bias 28, 29, 63, 98
   voltage 30–32, 35, 36, 39, 41, 47, 48, 59, 63, 65, 77, 94, 98, 111
approximations 34, 46, 47, 110
Au 49, 50, 56, 61, 63, 65, 96, 105, 106, 138, 141
   electrode 64, 114

B

background doping 31, 37, 42, 114, 117
band 9, 15, 90, 127, 128, 130
   model 48, 51, 53
   structure 9
bandgap 7, 9, 10, 14, 15, 27, 99, 126
barrier 20, 42, 63, 64, 131
   height 119, 120, 131
biexcitonic state 82, 83
bilayer structure 78
binomial 40
bipolaron 14, 26
black spots 87, 88, 90
blends 62, 63, 77, 99, 103, 111, 122
blocking layer 70, 72, 73, 76
Blodgett film 18
blue
   emitter 71, 79
   emitting host materials 74
   light 72, 78, 79
   OLEDs 71, 72, 78
   region 71, 72, 85
bond lengths 12
boundary conditions 30, 31, 33, 34, 36
breakdown 27, 67
brightness 78–80
buckminsterfullerene 24, 97
bulk heterojunction 63, 105, 111

C

Ca 61, 70, 77, 86, 96, 99, 100, 105
calculations 19, 22, 31, 35–37, 40–42, 45, 46, 56, 58, 60, 61, 119, 122, 127
carbon 2, 88, 101
carrier 17, 64, 76, 85, 88, 107, 116, 120, 126
   concentration 18, 120
   density 29, 31, 42, 49, 61, 128
   mobility 133, 136, 138
   traps 138

cathode 39, 51, 63, 64, 70, 83, 86–90, 92, 101, 106
cells 95, 96, 98, 100, 101, 105, 115, 117
channel 17, 18, 130, 133, 137
charge
  carrier transport mechanism 51
  carriers 4, 5, 13, 14, 16, 21, 48, 51, 59, 62, 72, 77, 85, 131, 140
    electrons 131
    injection 140
  transfer 77, 99, 103, 107
  transport 48, 51, 55, 72
    layer 77
charged solitons 13, 15, 21, 22
chemical structures 67, 69, 72, 75, 87, 91, 101, 104, 105, 107, 131, 133, 142, 144
chromaticity 71, 76
color tuning 71, 91
colors 5, 67, 71, 74, 75, 79, 81
complexes 91, 94
composition 76, 98, 108, 111, 112
computer simulations 64
conducting polymers 1–5, 7, 8, 12, 17, 18, 24, 27, 29, 60, 70, 97
conduction band 10, 13, 48, 128
  edge 14, 126, 128
conductivity 1–3, 20, 21, 27, 28, 103, 125
  high 3, 20
  of polymers 2, 4
conjugated polymers 1, 24, 70, 81, 82, 140–142
contact 28, 29, 36, 38, 57, 58, 63, 86, 94, 106, 111, 118, 138, 140
  injecting 39, 41, 42, 49
  layers 97, 133
  metal 106, 138, 140
  resistance 84, 123, 138
Contex 2
continuity equations 28–30, 39, 53
conversion 71, 75, 78
  efficiency 95, 97, 99
co-polymerization 75
copper 1, 3, 4
cost 95, 101, 123
crystallization 86
currents 49, 96, 108, 113, 114, 117, 121, 126
curves
  experimental 129
  illuminated 114, 115, 118

D
dark
  characteristics 111, 117
  currents 100, 112–114, 121
    measured 114, 115, 121
  powder 1
  saturation current 110, 111, 116
  spot 90
    areas 89
    growths 89
  transfer characteristics change 142
defects 5, 14, 18, 20, 85, 130
deformation 12, 14
degenerate 7, 8, 12, 13, 24, 25
degradation 83, 84, 86, 87, 90, 136
delamination 88
density of states 33, 56, 126, 130, 133
depths 46, 107, 108
detrapping 16, 17, 58
device
  architecture 104
  band diagram 63
  characteristics 144, 145
  design 138, 140
  lifetime 72
  operation 79, 140
  performance 73, 86, 91, 140, 144
    extrinsic 140
    improved 131
  processing 138
  properties 145
devices 1, 3–5, 51–53, 61–65, 67, 71–73, 75, 76, 78, 79, 81, 86–88, 90–95, 104, 114, 115, 136–138, 141–143
  alloy 104
  amorphous Si 123
  characteristics of 94
  emitting 70, 84
  inorganic 70
  liquid crystal display 67
  memory 1
  microcavity-structured 78
  optoelectronic 5, 67, 123
  organic molecular solid 67
  reference 92
  single mode cavity 79
  solution-processed 94
Dexter energy transfer 75, 77
dichroic ratio 136
dielectric 136, 141, 142, 144, 145

constant 27–29, 56
difference 12, 15, 42, 86, 91, 96, 98, 99, 104–106, 122, 126, 133
  work function 63, 104, 106
diffusion 18, 20, 77, 106, 107, 116, 117
  coefficient 16, 19, 20
  length 97, 119, 121, 122
dimerization 10, 11
dimmers 80
diodes 38, 41, 50, 61, 63, 80, 83, 96
dislocations 85
displacement 11–14
display 67, 68, 70–72, 79, 80, 124
  applications 74
displays 4, 5, 67, 68, 75, 79, 123, 124, 136
dissociation 23, 97–99, 114, 115
distortion 11, 12, 14, 24
domain size 108
donor 8, 13, 14, 21, 77, 96, 97, 99, 101, 104, 106
  polymers 99, 122
dopant 8, 14, 72, 74, 76, 79
  concentration 8, 20–22
  ions 8, 9, 14
doping 1, 7, 8, 13, 14, 22, 24, 29, 75, 115, 119
  concentration 14, 76, 77, 120
    background 29, 31, 32
down conversion 75, 78
drain 16, 17, 126, 128, 133
drift-diffusion transport model, conventional 134
dye molecules 70
dyes 70, 72, 79, 83, 95

E

efficiency 5, 67, 75, 81, 87, 96–101, 119, 121, 122
eigenvalues 11
electric field 23, 29, 31, 34, 36, 37, 41, 58, 60, 62, 98, 99, 107, 111, 126, 127, 130
electrochemical 8, 86
  cells 1, 8
electrode work function 105–107
electrodes 8, 23, 51, 68, 84, 86, 96, 100, 101, 104, 106, 107, 122
  material 122
electroluminescence 4, 5, 16, 62, 67, 78, 80, 100
electroluminescent materials, organic 75

electromagnetic shielding 2, 4
electron 9, 11–16, 23, 29, 31, 32, 51–53, 61, 63, 64, 70, 77, 83, 90, 97–99, 110
  acceptor 104, 105
  affinity 97, 98, 107
  devices 61
  injection voltage 86
  interactions 11, 14, 15
  mobility 27, 53, 62, 63
    intrinsic 53
    real intrinsic 53
  only device 52, 53, 61
  transport materials 51
electronic
  devices 5, 98
  structure 13
electron-injecting cathode, efficient 51
electrostatic 2, 62
emission 23, 72, 76–78, 81–83, 86, 91, 94
  colors 71, 75, 76
  spectrum 77–79
emissive layers 72, 75, 76, 79
encapsulation 84, 90, 122
energy 2, 10, 12, 13, 15, 16, 21, 24, 35, 47, 49, 55, 58, 59, 70, 71, 76, 77, 107, 127, 128
  levels 10, 13, 14
  space 32, 34, 45, 51, 60
  transfer 76, 77, 79
equations 11, 32–37, 39, 40, 44, 46, 47, 55, 57, 58, 60, 118, 121, 125–128
  diode 116, 117
excimer 75, 77
exciplex 75, 77
  formation 77
excitation 15, 16, 21, 70, 82, 83, 106
excited state 77, 84
excitons 3, 23, 77, 97, 98, 114, 115, 122
experimental
  data 38, 42, 44, 45, 48–53, 56, 61, 116–118, 129, 130, 138
  results 16, 17, 38, 51, 57, 58, 111, 114, 117, 124, 129
experiments 3, 4, 11, 14, 20, 22, 23, 42, 46, 58, 60, 61, 86, 87, 99, 110, 116, 117, 127, 129
exponential 35, 39, 42, 51, 55, 60, 64, 65, 116, 134, 136
expression 31, 32, 34–36, 119

F

Fabry–Perot 67, 78
Fermi
  energy 34
  function 59
  level 32, 64, 106, 119, 126, 130
Fermions 21
field effect mobility 123, 131, 133, 135, 136, 143, 145
films 1, 2, 7, 8, 70, 82, 86, 91, 104, 107, 108, 138
fit 20, 44, 45, 53, 57, 58, 103, 112, 116–118
fitting 48, 52, 53, 117, 129, 130
flat
  band 63
  panel 4, 5, 79, 123
flexible 1, 5, 95, 123, 145
flight method 16, 17
fluorescent 70, 81
fluoropolymer 144
formation 10, 72, 77, 80, 87, 90
Forster 75–77
  energy transfer 77
Fowler–Nordheim 64
free carrier density 37, 98, 128
Frenkel
  effect 58, 59
  value 58
full color display 70, 71, 79, 80, 124
fullerene 97, 98, 104, 141

G

gate 17, 128, 131
  dielectrics 141, 143
  voltage 126, 128, 129, 131, 134, 138, 142
    dependent mobility 137
    function of 124, 129
Gaussian
  distribution 32, 34, 56
  traps, shallow 55, 57
grain
  boundaries 119, 120, 130, 131, 138
  size 120, 121, 138
green 5, 67, 70, 71, 75, 79–81, 94
  emitter 71, 79
ground states 7, 12, 13, 24, 77, 103
growth 2, 7, 85, 143
guest 77

H

heterojunctions 115
heterostructure 27
hole 14, 23, 28, 29, 31, 46, 48, 51, 56, 57, 61–65, 70, 72, 73, 83, 92, 97, 98, 111
  blocking 70, 72, 73
  mobility 62, 63
  only device 48, 64, 65
  traps 32, 56
hopping 16, 53, 55, 126
host 77
8-hydroxyquinoline 51, 67, 91

I

illuminated
  characteristics 96, 104, 106, 111, 114, 115, 118, 121
  currents 113–115
illumination 62, 63, 74, 87, 98, 104, 114–119, 121, 122, 142, 143
  material parameters 110
improvement 72, 94, 101, 104, 123, 144, 145
impurities 14, 31, 33, 84
impurity 33, 61
incandescent lamp 74, 75, 81
inflexion 112, 114, 117
injected
  carrier density 36, 115
  carriers 32, 33
injection barrier 42, 44, 45
ink-jet printing 1, 72, 83
inorganic 1, 5, 23, 32, 70, 75, 78, 80, 83–85, 95, 125, 130
insulators 1, 9, 16, 28, 38, 103, 128, 138
interface 77, 87, 97, 107, 127, 144, 145
interpenetrating network 97, 98, 122
iodine 1, 20, 21
ion-beam 136, 137

J

junction 62, 63, 67, 70, 97, 104, 115–118, 122

L

large polymer screens 68
lasers 1, 5, 67, 80, 81, 83
lasing 82, 83
lattice 9, 13, 85
layer structure, schematic 87
layers 2, 9, 62, 68, 70, 76, 78, 85, 88, 96, 100, 101, 131, 138, 143

active 1, 44, 85, 87, 90, 92, 96–98, 100, 101, 104, 107, 122, 124, 126–128, 131, 133
  blocking 70, 72, 73, 76
  electron transporting 76, 83
  emitter 76, 77
  hole transporting 70, 76, 84
  mixed 99
  phosphor 78
LEDs 1, 3–5, 67, 71, 79, 80, 86, 90
  organic 75, 78, 84, 86–88, 94, 101
level, single energy 32, 57, 59, 60, 131
LiF 51, 52, 63, 88, 92, 94, 97, 101, 106, 114, 142, 143
  layer 101
lifetime 2, 72, 79, 81, 85–87, 90
  operational 86, 87, 90
light 4, 16, 17, 67, 70, 71, 74, 75, 78, 86, 87, 104, 106, 115, 116, 119, 120, 122, 123
  emission 67, 70, 86
  incident 23, 97, 99, 106
  intensity of 71, 121
limited currents 16, 32, 62, 112, 114, 115
localized states 55, 133, 136
low temperature activation energy 54
luminance 72, 74, 78, 89, 94
luminous efficiency 72, 74, 87, 94

M
mass 13
materials, organic 4, 5, 32, 37, 58, 72, 81, 82, 84, 85, 90, 91, 115, 130
matrix, active 123, 124
maximum luminance 72, 74, 77, 80
measured
  mobilities 144
  open-circuit voltages 106
measurements 15–18, 31, 51, 53, 61, 62, 90, 112, 114, 115, 118
metal electrodes 95, 105, 106
metallic mirrors 78
metals 2, 9, 22, 28, 29, 38, 86, 87, 105
microactuators 2
microcavities 67, 68, 78, 82
microcavity structure 75, 78
microwaves 2
midgap 12, 14, 24
  levels 14, 15
  states 14, 24
mirror symmetry 24

mixed host structure 72
mobility 5, 16–18, 20, 27, 49, 51, 55–57, 62, 63, 85, 86, 110, 123, 124, 130, 131, 136, 138, 144, 145
  anisotropic 136
  low 24, 85
  model 5, 46–48, 53, 55, 58, 61
  values 18, 67, 120, 145
molecular 2, 3, 9, 24, 51, 55, 62, 67, 68, 71, 72, 75, 76, 79, 85, 92, 106, 108, 130, 131
monolayer 2, 9, 138, 143
  thickness control 2, 9
morphology 86, 104, 107, 138, 139, 144
Mott's law 30
multilayer
  device structure 76
  devices 76–78
  structure 76
multimode resonant cavities 78

N
nanotube 101, 103
  concentration 103
nitrogen 85, 87
non-emitting dark spots 89
numerical calculations 37–39, 49, 60, 61, 127, 130, 131

O
ohmic contact 51
Ohm's law 28, 30, 32, 37, 42
OLEDs 4, 51, 67, 70–72, 76, 78, 79, 89–91, 93, 123
  white 71, 78–80
oligomers 5, 74, 83, 85
open circuit voltage 96, 100, 101, 104–107, 111, 113, 114, 119, 121, 122
optical
  absorption 14, 16, 26, 85, 87
  feedback 82
optically pumped 5, 68, 71, 81, 83, 94
optoelectronic 5, 67, 81
orbitals 9, 10
organic
  devices 28, 95
  electroluminescence 5
  lasers, pumped 83, 94
  semiconductor devices 123
  solar cells 5, 95, 98, 105, 114–118, 122

organics 39, 65, 67, 96, 102, 140
organometallic 73
output characteristics 112, 115, 117, 122, 133, 134, 141, 143
overlapping radiation fields 82
oxidation 86, 87
oxygen 86, 88, 90

P

packaging 2
parameters 11, 20, 31, 32, 36, 38, 40–42, 44, 45, 47, 49, 51–53, 56, 60, 61, 79, 116, 117, 128–130
  values of 11, 53
para-phenylene 24, 26, 27
Pd 106, 140, 141
pentacene 115, 133, 136, 138
  devices 133, 143, 144
  films 136, 138, 143, 145
  transistors 5, 144, 145
percolation 22, 103
performance 2, 5, 65, 74, 80, 90, 96–98, 101, 104, 123, 131, 133, 140, 143
perylene 72, 74, 83, 84
phase 8, 9, 108, 109
phonon 11, 19
phosphorescent 74
phosphors 75, 78, 81
photoconductivity 15, 16, 24, 97, 108
photocurrents 97, 99, 100
photodetectors 97
photodiodes 1, 97
photoemission 14, 24
photoexcitation 13
photoinduced changes 14–16
photoluminescence 23, 26, 78, 81, 87, 99
  polarized 136
photons 15, 16, 97–99
  incident 15
photooxidation 87
photoproduction 15, 21
photopumping 82
photostability 142
phototransistor 140, 142
photovoltaic 65, 95, 97, 102, 104, 118
pinholes 57, 87
plasma 28, 85, 90, 92
plastics 1, 2
Poisson equation 31, 32, 36, 39
polarons 3, 10, 12–14
  charged 13
  positive 14
polyacetylene 1, 3, 7, 22–24
polyaniline 2, 3, 8, 27
polycrystalline 119, 120, 130, 133, 138
polymer
  acceptor 101
  backbone 72, 85
  blend 23
  cells 111
  chain 13, 24
  concentration 86
  contact 86
  devices 68
  dielectrics 144, 145
  diodes 126, 128
  electromechanical mechanisms 2
  films 8, 72, 82, 86, 99
  laser diodes 82
  layer 95
    emissive 90
  light emitting diodes 65, 72
  medium, active 82
  optoelectronic devices 81
  photovoltaic devices 65
  solar cells 110, 116
polymeric matrices 72
polymers 1–4, 7–9, 13, 14, 16–18, 22–25, 27, 32, 59, 60, 64, 65, 67, 68, 72, 77, 85–89, 96–101, 125, 126
  doped 15, 28, 83
  electron-accepting 106
  excited 82
  hole-accepting 106
  interpenetrating 97
  oldest synthetic 3
  organic 1, 83
  precursor 8
  pristine conducting 3, 16
  semiconducting 81
  styrenic 144, 145
  undoped 14
polymethylmethacrylate 133
poly(p-phenylene vinylene) 17, 65
polypyrrole 8, 24, 27
polythiophene 8, 18, 27
power
  density 87

efficiency 73, 74, 76
law 39, 40, 42, 43, 48, 52
precursor 8

Q

quantum
  efficiency 72, 78, 83, 84, 100
    external 74, 80, 101, 115
  well 85
  yield 78
quenching 97, 99, 100, 108

R

recombination 70, 76, 116, 117
rectification 100
red 5, 71, 74, 75, 77–80
  emitter 71, 79
  light 70, 75, 78, 79, 81
regions
  disordered 27, 28
  high
    temperature 46
    voltage 49
  linear 54, 123, 136
  low temperature 54
resonance 16, 18, 20, 91
responsivity 140, 141
reverse bias 96, 98–100, 104, 111
room temperature 15, 46, 51, 53, 88, 115, 138, 139
  conductivity 22

S

sapphire 85
scanning electron microscopy 136
scattering 19, 28, 78
Schottky
  barrier 39, 40, 44, 48, 49, 57, 58, 64
  diodes 43, 95, 96, 122
SCLC 28, 32, 33, 37, 42, 48, 63, 115–118
semiconductor devices 1
semiconductors 9, 13, 14, 22, 23, 29, 85, 128, 144
  organic 4, 46, 51, 74, 83, 95, 123, 126, 131
sensors 123
series resistance 63, 110, 116
sexithiophene 18, 124, 131
  structure of 131, 132
shallow traps 57, 130

short circuit 96, 100, 101, 104, 108, 110, 111, 114, 116, 121
single
  layer 75, 79, 87, 114
    devices 87, 96, 105
    structure 79
  molecules 79, 80, 85
singlets 77, 84
SiO_2 143, 144
  native 136
Si solar cells 95, 98, 115, 116
small molecules 70, 74, 77, 97, 122
smart cards 5, 123
solar cells 1, 3, 5, 95–98, 100, 101, 108, 111, 112, 115–119, 121, 122
  amorphous 116
  bilayer 95–97, 106, 122
  bulk heterojunction 5, 63, 98, 101
  conducting polymer 95
  illuminated 116, 117
    organic 115–117
  single layer 95
solid state lasers 81
  organic 81, 82
soliton–antisoliton pair 15, 24, 26
soliton energy levels 14
solitons 10, 12–15, 19–22, 24
source 17, 38, 57, 74, 75, 83, 94, 95, 104, 126, 127, 133, 137, 138, 140–143, 145
space charge 32, 42, 48, 62–64, 112, 114–116, 119–122, 126
  layer 110
  limited currents 28
spectra 71, 79, 81, 82, 97
  parallel electron-energy-loss 98
  photocurrent action 96
spectral narrowing 78, 82, 83
spectroscopy 14, 62, 87, 91, 136
spectrum 24, 71, 74, 77–79, 81, 82, 91, 100, 101, 104, 122
spin coating 1, 72, 80, 83, 98, 104
spins 13, 14, 21, 22, 86, 100
  zero 21, 22
spontaneous emission 78, 82
spots 87, 88
  hot 87
stability 86, 87, 90, 131
stacked 75, 79, 80
statistics 13
stimulated emission 70, 81–83

stretching 3, 8
substrates 38, 82, 85, 91, 92, 123, 133, 136, 143
superfluorescence 82
superposition principle 108, 110–112, 116
superradiance 82
surface 17, 67, 89, 90, 108, 126–129, 136
　voltage 128
susceptibility 20, 22, 24

T

temperature
　behavior 130
　dependence 17, 57
　effects 44, 50, 51
　range 48, 49, 134
temperatures 5, 19, 20, 22, 31, 32, 39, 44–46, 48–54, 63, 74, 80, 118, 119, 126, 130, 131, 133–136, 138
　high 8, 46, 53–55, 90
　low 15, 28, 39, 46, 49, 51, 53–55, 65, 85, 126, 138
theory 3, 4, 11, 14, 15, 20, 38, 39, 41, 42, 44, 46, 49, 51, 53, 57, 58, 115–117, 126, 127, 129, 130
　new 42, 44, 50, 51
thermal evaporation 90, 104, 133
thickness 2, 9, 17, 29, 30, 36, 44, 45, 47, 51, 52, 61, 63, 64, 70, 76, 78, 86–88, 107, 108
thin films 3, 22, 67, 82, 91, 130, 131, 136
　organic 81, 82
thiophene 26, 72, 77, 80
threshold 15, 16, 21, 83
　power 70, 83
　voltage 86, 125, 131, 133, 135, 138, 144
　　near zero 123
top electrode 63
transfer 11, 70, 97, 98, 134
　characteristics 133–136, 143
　energy 79
transient 14, 16, 62
transistors 5, 123, 124, 126, 131, 138, 140, 143–145
　amorphous Si 124
　channel region of 145
　organic thin film 5, 67, 123, 127
　thin film 16, 124, 144
transition 12, 15, 22, 42
　band-to-band 14

transmission 14–16, 79
trans-polyacetylene 7, 9, 27
transport 5, 16, 31, 38, 39, 51, 53, 55, 58, 64, 125, 130, 131
　equations 32–34, 59
　layer 77
　mechanisms 123
　properties 4, 16, 23, 51, 65, 98, 122
trap
　density 33, 48, 126, 130
　depth 32, 34, 58, 59, 126, 127
　levels 32, 128, 130
trap-filled limit 33, 36
trap-free 30, 33, 36, 37
　insulators 28, 29, 116
trapped carrier density 127, 128, 131
trapping 16, 17
　model 44–46, 51, 53, 58
traps 32–36, 46, 49, 51, 53, 55, 57–60, 64, 98, 115–117, 119, 120, 125–127, 130, 131, 145
　distributed 32, 34, 35, 45, 47, 56–60
　distribution of 34, 51, 138
　exponential distribution of 51, 55, 64
　limited 57
　multiple 138
　shallow 57, 130
　single
　　energy level 32
　　level 32, 55, 60, 61, 128
treatment 28, 29, 31, 90
triplet 74, 77
tunnel 46
tunneling 48, 77, 126, 131

U

unified model 55, 57

V

vacuum 74, 87, 91, 135
valence band 33, 126
values 11, 13, 14, 17–20, 27–32, 34, 35, 39–42, 44–47, 49, 53, 54, 56, 58–61, 96–98, 100, 116, 117, 129–131
　bandgap 101, 122
　best 145
　calculated 12, 19, 29, 37, 41, 44, 45, 47, 49, 56, 121, 129, 131
　expected 114
　experimental 12, 16, 54, 56

fixed 127
large 37, 42
measured 20, 131
observed 16
reliable 17, 58, 117
reported 49
reproducible 127
small 28, 41, 42
voltage 32, 96, 105, 111
  measurements 62
  range 114
    intermediate 51
  tunable performances 71
voltages 17, 28, 31–33, 36, 37, 41–44, 48, 49, 62, 63, 65, 86, 94, 106, 107, 111, 112, 114–116, 118, 128
  built-in 63, 118
  calculated 38
  of deviation 38, 42
  emitting 86
  external 101
  high 36, 46, 49, 65
  intermediate 48
  low 29, 33, 42, 48, 51, 57
  small 58, 115, 117, 118
  total 59
  turn-on 98

W

water vapors 86, 87, 90
waveguide 82
white light 71, 74–78, 80, 81
  emission 70, 74–80
work
  function values 96, 122
  functions 90, 96, 101, 105–107, 122
  negative electrode 105

X

X-ray photoelectron spectroscopy 87

Z

zero field mobility 57

Contents of Volumes in This Series

Volume 1 Physics of III–V Compounds

C. Hilsum, Some Key Features of III–V Compounds
F. Bassani, Methods of Band Calculations Applicable to III–V Compounds
E. O. Kane, The $k \cdot p$ Method
V. L. Bonch-Bruevich, Effect of Heavy Doping on the Semiconductor Band Structure
D. Long, Energy Band Structures of Mixed Crystals of III–V Compounds
L. M. Roth and P. N. Argyres, Magnetic Quantum Effects
S. M. Puri and T. H. Geballe, Thermomagnetic Effects in the Quantum Region
W. M. Becker, Band Characteristics near Principal Minima from Magnetoresistance
E. H. Putley, Freeze-Out Effects, Hot Electron Effects, and Submillimeter Photoconductivity in InSb
H. Weiss, Magnetoresistance
B. Ancker-Johnson, Plasma in Semiconductors and Semimetals

Volume 2 Physics of III–V Compounds

M. G. Holland, Thermal Conductivity
S. I. Novkova, Thermal Expansion
U. Piesbergen, Heat Capacity and Debye Temperatures
G. Giesecke, Lattice Constants
J. R. Drabble, Elastic Properties
A. U. Mac Rae and G. W. Gobeli, Low Energy Electron Diffraction Studies
R. Lee Mieher, Nuclear Magnetic Resonance
B. Goldstein, Electron Paramagnetic Resonance
T. S. Moss, Photoconduction in III–V Compounds
E. Antoncik and J. Tauc, Quantum Efficiency of the Internal Photoelectric Effect in InSb
G. W. Gobeli and I. G. Allen, Photoelectric Threshold and Work Function
P. S. Pershan, Nonlinear Optics in III–V Compounds
M. Gershenzon, Radiative Recombination in the III–V Compounds
F. Stern, Stimulated Emission in Semiconductors

Volume 3 Optical Properties of III–V Compounds

M. Hass, Lattice Reflection
W. G. Spitzer, Multiphonon Lattice Absorption

D. L. Stierwalt and R. F. Potter, Emittance Studies

H. R. Philipp and H. Ehrenveich, Ultraviolet Optical Properties

M. Cardona, Optical Absorption Above the Fundamental Edge

E. J. Johnson, Absorption Near the Fundamental Edge

J. O. Dimmock, Introduction to the Theory of Exciton States in Semiconductors

B. Lax and J. G. Mavroides, Interband Magnetooptical Effects

H. Y. Fan, Effects of Free Carries on Optical Properties

E. D. Palik and G. B. Wright, Free-Carrier Magnetooptical Effects

R. H. Bube, Photoelectronic Analysis

B. O. Seraphin and H. E. Benett, Optical Constants

Volume 4 Physics of III–V Compounds

N. A. Goryunova, A. S. Borchevskii and D. N. Tretiakov, Hardness

N. N. Sirota, Heats of Formation and Temperatures and Heats of Fusion of Compounds of $A^{III}B^{V}$

D. L. Kendall, Diffusion

A. G. Chynoweth, Charge Multiplication Phenomena

R. W. Keyes, The Effects of Hydrostatic Pressure on the Properties of III–V Semiconductors

L. W. Aukerman, Radiation Effects

N. A. Goryunova, F. P. Kesamanly, and D. N. Nasledov, Phenomena in Solid Solutions

R. T. Bate, Electrical Properties of Nonuniform Crystals

Volume 5 Infrared Detectors

H. Levinstein, Characterization of Infrared Detectors

P. W. Kruse, Indium Antimonide Photoconductive and Photoelectromagnetic Detectors

M. B. Prince, Narrowband Self-Filtering Detectors

I. Melngalis and T. C. Hannan, Single-Crystal Lead-Tin Chalcogenides

D. Long and J. L. Schmidt, Mercury-Cadmium Telluride and Closely Related Alloys

E. H. Putley, The Pyroelectric Detector

N. B. Stevens, Radiation Thermopiles

R. J. Keyes and T. M. Quist, Low Level Coherent and Incoherent Detection in the Infrared

M. C. Teich, Coherent Detection in the Infrared

F. R. Arams, E. W. Sard, B. J. Peyton and F. P. Pace, Infrared Heterodyne Detection with Gigahertz IF Response

H. S. Sommers, Jr., Macrowave-Based Photoconductive Detector

R. Sehr and R. Zuleeg, Imaging and Display

Volume 6 Injection Phenomena

M. A. Lampert and R. B. Schilling, Current Injection in Solids: The Regional Approximation Method

R. Williams, Injection by Internal Photoemission

A. M. Barnett, Current Filament Formation

R. Baron and J. W. Mayer, Double Injection in Semiconductors

W. Ruppel, The Photoconductor-Metal Contact

Volume 7 Application and Devices

Part A

J. A. Copeland and S. Knight, Applications Utilizing Bulk Negative Resistance
F. A. Padovani, The Voltage-Current Characteristics of Metal-Semiconductor Contacts
P. L. Hower, W. W. Hooper, B. R. Cairns, R. D. Fairman, and D. A. Tremere, The GaAs Field-Effect Transistor
M. H. White, MOS Transistors
G. R. Antell, Gallium Arsenide Transistors
T. L. Tansley, Heterojunction Properties

Part B

T. Misawa, IMPATT Diodes
H. C. Okean, Tunnel Diodes
R. B. Campbell and Hung-Chi Chang, Silicon Junction Carbide Devices
R. E. Enstrom, H. Kressel, and L. Krassner, High-Temperature Power Rectifiers of $GaAs_{1-x}P_x$

Volume 8 Transport and Optical Phenomena

R. J. Stirn, Band Structure and Galvanomagnetic Effects in III–V Compounds with Indirect Band Gaps
R. W. Ure, Jr., Thermoelectric Effects in III–V Compounds
H. Piller, Faraday Rotation
H. Barry Bebb and E. W. Williams, Photoluminescence I: Theory
E. W. Williams and H. Barry Bebb, Photoluminescence II: Gallium Arsenide

Volume 9 Modulation Techniques

B. O. Seraphin, Electroreflectance
R. L. Aggarwal, Modulated Interband Magnetooptics
D. F. Blossey and Paul Handler, Electroabsorption
B. Batz, Thermal and Wavelength Modulation Spectroscopy
I. Balslev, Piezooptical Effects
D. E. Aspnes and N. Bottka, Electric-Field Effects on the Dielectric Function of Semiconductors and Insulators

Volume 10 Transport Phenomena

R. L. Rhode, Low-Field Electron Transport
J. D. Wiley, Mobility of Holes in III–V Compounds
C. M. Wolfe and G. E. Stillman, Apparent Mobility Enhancement in Inhomogeneous Crystals
R. L. Petersen, The Magnetophonon Effect

Volume 11 Solar Cells

H. J. Hovel, Introduction; Carrier Collection, Spectral Response, and Photocurrent; Solar Cell Electrical Characteristics; Efficiency; Thickness; Other Solar Cell Devices; Radiation Effects; Temperature and Intensity; Solar Cell Technology

Volume 12 Infrared Detectors (II)

W. L. Eiseman, J. D. Merriam, and R. F. Potter, Operational Characteristics of Infrared Photodetectors
P. R. Bratt, Impurity Germanium and Silicon Infrared Detectors
E. H. Putley, InSb Submillimeter Photoconductive Detectors
G. E. Stillman, C. M. Wolfe, and J. O. Dimmock, Far-Infrared Photoconductivity in High Purity GaAs
G. E. Stillman and C. M. Wolfe, Avalanche Photodiodes
P. L. Richards, The Josephson Junction as a Detector of Microwave and Far-Infrared Radiation
E. H. Putley, The Pyroelectric Detector – An Update

Volume 13 Cadmium Telluride

K. Zanio, Materials Preparations; Physics; Defects; Applications

Volume 14 Lasers, Junctions, Transport

N. Holonyak, Jr., and M. H. Lee, Photopumped III–V Semiconductor Lasers
H. Kressel and J. K. Butler, Heterojunction Laser Diodes
A. Van der Ziel, Space-Charge-Limited Solid-State Diodes
P. J. Price, Monte Carlo Calculation of Electron Transport in Solids

Volume 15 Contacts, Junctions, Emitters

B. L. Sharma, Ohmic Contacts to III–V Compounds Semiconductors
A. Nussbaum, The Theory of Semiconducting Junctions
J. S. Escher, NEA Semiconductor Photoemitters

Volume 16 Defects, (HgCd)Se, (HgCd)Te

H. Kressel, The Effect of Crystal Defects on Optoelectronic Devices
C. R. Whitsett, J. G. Broerman, and C. J. Summers, Crystal Growth and Properties of $Hg_{1-x}Cd_x$ Se Alloys
M. H. Weiler, Magnetooptical Properties of $Hg_{1-x}Cd_x$ Te Alloys
P. W. Kruse and J. G. Ready, Nonlinear Optical Effects in $Hg_{1-x}Cd_x$ Te

Volume 17 CW Processing of Silicon and Other Semiconductors

J. F. Gibbons, Beam Processing of Silicon
A. Lietoila, R. B. Gold, J. F. Gibbons, and L. A. Christel, Temperature Distributions and Solid Phase Reaction Rates Produced by Scanning CW Beams
A. Leitoila and J. F. Gibbons, Applications of CW Beam Processing to Ion Implanted Crystalline Silicon
N. M. Johnson, Electronic Defects in CW Transient Thermal Processed Silicon
K. F. Lee, T. J. Stultz, and J. F. Gibbons, Beam Recrystallized Polycrystalline Silicon: Properties, Applications, and Techniques
T. Shibata, A. Wakita, T. W. Sigmon and J. F. Gibbons, Metal-Silicon Reactions and Silicide
Y. I. Nissim and J. F. Gibbons, CW Beam Processing of Gallium Arsenide

Volume 18 Mercury Cadmium Telluride

P. W. Kruse, The Emergence of $(Hg_{1-x} Cd_x)Te$ as a Modern Infrared Sensitive Material
H. E. Hirsch, S. C. Liang, and A. G. White, Preparation of High-Purity Cadmium, Mercury, and Tellurium
W. F. H. Micklethwaite, The Crystal Growth of Cadmium Mercury Telluride
P. E. Petersen, Auger Recombination in Mercury Cadmium Telluride
R. M. Broudy and V. J. Mazurczyck, (HgCd)Te Photoconductive Detectors
M. B. Reine, A. K. Soad, and T. J. Tredwell, Photovoltaic Infrared Detectors
M. A. Kinch, Metal-Insulator-Semiconductor Infrared Detectors

Volume 19 Deep Levels, GaAs, Alloys, Photochemistry

G. F. Neumark and K. Kosai, Deep Levels in Wide Band-Gap III–V Semiconductors
D. C. Look, The Electrical and Photoelectronic Properties of Semi-Insulating GaAs
R. F. Brebrick, Ching-Hua Su, and Pok-Kai Liao, Associated Solution Model for Ga-In-Sb and Hg-Cd-Te
Y. Ya. Gurevich and Y. V. Pleskon, Photoelectrochemistry of Semiconductors

Volume 20 Semi-Insulating GaAs

R. N. Thomas, H. M. Hobgood, G. W. Eldridge, D. L. Barrett, T. T. Braggins, L. B. Ta, and S. K. Wang, High-Purity LEC Growth and Direct Implantation of GaAs for Monolithic Microwave Circuits
C. A. Stolte, Ion Implantation and Materials for GaAs Integrated Circuits
C. G. Kirkpatrick, R. T. Chen, D. E. Holmes, P. M. Asbeck, K. R. Elliott, R. D. Fairman, and J. R. Oliver, LEC GaAs for Integrated Circuit Applications
J. S. Blakemore and S. Rahimi, Models for Mid-Gap Centers in Gallium Arsenide

Volume 21 Hydrogenated Amorphous Silicon

Part A
J. I. Pankove, Introduction
M. Hirose, Glow Discharge; Chemical Vapor Deposition
Y. Uchida, di Glow Discharge
T. D. Moustakas, Sputtering
I. Yamada, Ionized-Cluster Beam Deposition
B. A. Scott, Homogeneous Chemical Vapor Deposition
F. J. Kampas, Chemical Reactions in Plasma Deposition
P. A. Longeway, Plasma Kinetics
H. A. Weakliem, Diagnostics of Silane Glow Discharges Using Probes and Mass Spectroscopy
L. Gluttman, Relation between the Atomic and the Electronic Structures
A. Chenevas-Paule, Experiment Determination of Structure
S. Minomura, Pressure Effects on the Local Atomic Structure
D. Adler, Defects and Density of Localized States

Part B
J. I. Pankove, Introduction
G. D. Cody, The Optical Absorption Edge of a-Si: H
N. M. Amer and W. B. Jackson, Optical Properties of Defect States in a-Si: H

P. J. Zanzucchi, The Vibrational Spectra of a-Si: H
Y. Hamakawa, Electroreflectance and Electroabsorption
J. S. Lannin, Raman Scattering of Amorphous Si, Ge, and Their Alloys
R. A. Street, Luminescence in a-Si: H
R. S. Crandall, Photoconductivity
J. Tauc, Time-Resolved Spectroscopy of Electronic Relaxation Processes
P. E. Vanier, IR-Induced Quenching and Enhancement of Photoconductivity and Photoluminescence
H. Schade, Irradiation-Induced Metastable Effects
L. Ley, Photoelectron Emission Studies

Part C

J. I. Pankove, Introduction
J. D. Cohen, Density of States from Junction Measurements in Hydrogenated Amorphous Silicon
P. C. Taylor, Magnetic Resonance Measurements in a-Si: H
K. Morigaki, Optically Detected Magnetic Resonance
J. Dresner, Carrier Mobility in a-Si: H
T. Tiedje, Information About Band-Tail States from Time-of-Flight Experiments
A. R. Moore, Diffusion Length in Undoped a-S: H
W. Beyer and J. Overhof, Doping Effects in a-Si: H
H. Fritzche, Electronic Properties of Surfaces in a-Si: H
C. R. Wronski, The Staebler-Wronski Effect
R. J. Nemanich, Schottky Barriers on a-Si: H
B. Abeles and T. Tiedje, Amorphous Semiconductor Superlattices

Part D

J. I. Pankove, Introduction
D. E. Carlson, Solar Cells
G. A. Swartz, Closed-Form Solution of I–V Characteristic for a s-Si: H Solar Cells
I. Shimizu, Electrophotography
S. Ishioka, Image Pickup Tubes
P. G. Lecomber and W. E. Spear, The Development of the a-Si: H Field-Effect Transistor and its Possible Applications
D. G. Ast, a-Si:H FET-Addressed LCD Panel
S. Kaneko, Solid-State Image Sensor
M. Matsumura, Charge-Coupled Devices
M. A. Bosch, Optical Recording
A. D'Amico and G. Fortunato, Ambient Sensors
H. Kulkimoto, Amorphous Light-Emitting Devices
R. J. Phelan, Jr., Fast Decorators and Modulators
J. I. Pankove, Hybrid Structures
P. G. LeComber, A. E. Owen, W. E. Spear, J. Hajto, and W. K. Choi, Electronic Switching in Amorphous Silicon Junction Devices

Volume 22 Lightwave Communications Technology

Part A

K. Nakajima, The Liquid-Phase Epitaxial Growth of InGaAsP

W. T. Tsang, Molecular Beam Epitaxy for III–V Compound Semiconductors
G. B. Stringfellow, Organometallic Vapor-Phase Epitaxial Growth of III–V Semiconductors
G. Beuchet, Halide and Chloride Transport Vapor-Phase Deposition of InGaAsP and GaAs
M. Razeghi, Low-Pressure, Metallo-Organic Chemical Vapor Deposition of $Ga_xIn_{1-x}AsP_{1-y}$ Alloys
P. M. Petroff, Defects in III–V Compound Semiconductors

Part B
J. P. van der Ziel, Mode Locking of Semiconductor Lasers
K. Y. Lau and A. Yariv, High-Frequency Current Modulation of Semiconductor Injection Lasers
C. H. Henry, Special Properties of Semi Conductor Lasers
Y. Suematsu, K. Kishino, S. Arai, and F. Koyama, Dynamic Single-Mode Semiconductor Lasers with a Distributed Reflector
W. T. Tsang, The Cleaved-Coupled-Cavity (C^3) Laser

Part C
R. J. Nelson and N. K. Dutta, Review of InGaAsP InP Laser Structures and Comparison of Their Performance
N. Chinone and M. Nakamura, Mode-Stabilized Semiconductor Lasers for 0.7–0.8- and 1.1–1.6-μm Regions
Y. Horikoshi, Semiconductor Lasers with Wavelengths Exceeding 2 μm
B. A. Dean and M. Dixon, The Functional Reliability of Semiconductor Lasers as Optical Transmitters
R. H. Saul, T. P. Lee, and C. A. Burus, Light-Emitting Device Design
C. L. Zipfel, Light-Emitting Diode-Reliability
T. P. Lee and T. Li, LED-Based Multimode Lightwave Systems
K. Ogawa, Semiconductor Noise-Mode Partition Noise

Part D
F. Capasso, The Physics of Avalanche Photodiodes
T. P. Pearsall and M. A. Pollack, Compound Semiconductor Photodiodes
T. Kaneda, Silicon and Germanium Avalanche Photodiodes
S. R. Forrest, Sensitivity of Avalanche Photodetector Receivers for High-Bit-Rate Long-Wavelength Optical Communication Systems
J. C. Campbell, Phototransistors for Lightwave Communications

Part E
S. Wang, Principles and Characteristics of Integrable Active and Passive Optical Devices
S. Margalit and A. Yariv, Integrated Electronic and Photonic Devices
T. Mukai, Y. Yamamoto, and T. Kimura, Optical Amplification by Semiconductor Lasers

Volume 23 Pulsed Laser Processing of Semiconductors

R. F. Wood, C. W. White and R. T. Young, Laser Processing of Semiconductors: An Overview
C. W. White, Segregation, Solute Trapping and Supersaturated Alloys
G. E. Jellison, Jr., Optical and Electrical Properties of Pulsed Laser-Annealed Silicon
R. F. Wood and G. E. Jellison, Jr., Melting Model of Pulsed Laser Processing
R. F. Wood and F. W. Young, Jr., Nonequilibrium Solidification Following Pulsed Laser Melting
D. H. Lawndes and G. E. Jellison, Jr., Time-Resolved Measurement During Pulsed Laser Irradiation of Silicon
D. M. Zebner, Surface Studies of Pulsed Laser Irradiated Semiconductors
D. H. Lowndes, Pulsed Beam Processing of Gallium Arsenide

R. B. James, Pulsed CO_2 Laser Annealing of Semiconductors

R. T. Young and R. F. Wood, Applications of Pulsed Laser Processing

Volume 24 Applications of Multiquantum Wells, Selective Doping, and Superlattices

C. Weisbuch, Fundamental Properties of III–V Semiconductor Two-Dimensional Quantized Structures: The Basis for Optical and Electronic Device Applications

H. Morkoç and H. Unlu, Factors Affecting the Performance of (Al,Ga)As/GaAs and (Al,Ga)As/InGaAs Modulation-Doped Field-Effect Transistors: Microwave and Digital Applications

N. T. Linh, Two-Dimensional Electron Gas FETs: Microwave Applications

M. Abe et al., Ultra-High-Speed HEMT Integrated Circuits

D. S. Chemla, D. A. B. Miller and P. W. Smith, Nonlinear Optical Properties of Multiple Quantum Well Structures for Optical Signal Processing

F. Capasso, Graded-Gap and Superlattice Devices by Band-Gap Engineering

W. T. Tsang, Quantum Confinement Heterostructure Semiconductor Lasers

G. C. Osbourn et al., Principles and Applications of Semiconductor Strained-Layer Superlattices

Volume 25 Diluted Magnetic Semiconductors

W. Giriat and J. K. Furdyna, Crystal Structure, Composition, and Materials Preparation of Diluted Magnetic Semiconductors

W. M. Becker, Band Structure and Optical Properties of Wide-Gap $A_{1-x}^{II}Mn_xB_{IV}$ Alloys at Zero Magnetic Field

S. Oseroff and P. H. Keesom, Magnetic Properties: Macroscopic Studies

T. Giebultowicz and T. M. Holden, Neutron Scattering Studies of the Magnetic Structure and Dynamics of Diluted Magnetic Semiconductors

J. Kossut, Band Structure and Quantum Transport Phenomena in Narrow-Gap Diluted Magnetic Semiconductors

C. Riquaux, Magnetooptical Properties of Large-Gap Diluted Magnetic Semiconductors

J. A. Gaj, Magnetooptical Properties of Large-Gap Diluted Magnetic Semiconductors

J. Mycielski, Shallow Acceptors in Diluted Magnetic Semiconductors: Splitting, Boil-off, Giant Negative Magnetoresistance

A. K. Ramadas and R. Rodriquez, Raman Scattering in Diluted Magnetic Semiconductors

P. A. Wolff, Theory of Bound Magnetic Polarons in Semimagnetic Semiconductors

Volume 26 III–V Compound Semiconductors and Semiconductor Properties of Superionic Materials

Z. Yuanxi, III–V Compounds

H. V. Winston, A. T. Hunter, H. Kimura, and R. E. Lee, InAs-Alloyed GaAs Substrates for Direct Implantation

P. K. Bhattacharya and S. Dhar, Deep Levels in III–V Compound Semiconductors Grown by MBE

Y. Ya. Gurevich and A. K. Ivanov-Shits, Semiconductor Properties of Supersonic Materials

Volume 27 High Conducting Quasi-One-Dimensional Organic Crystals

E. M. Conwell, Introduction to Highly Conducting Quasi-One-Dimensional Organic Crystals

I. A. Howard, A Reference Guide to the Conducting Quasi-One-Dimensional Organic Molecular Crystals
J. P. Pouqnet, Structural Instabilities
E. M. Conwell, Transport Properties
C. S. Jacobsen, Optical Properties
J. C. Scolt, Magnetic Properties
L. Zuppiroli, Irradiation Effects: Perfect Crystals and Real Crystals

Volume 28 Measurement of High-Speed Signals in Solid State Devices

J. Frey and D. Ioannou, Materials and Devices for High-Speed and Optoelectronic Applications
H. Schumacher and E. Strid, Electronic Wafer Probing Techniques
D. H. Auston, Picosecond Photoconductivity: High-Speed Measurements of Devices and Materials
J. A. Valdmanis, Electro-Optic Measurement Techniques for Picosecond Materials, Devices and Integrated Circuits
J. M. Wiesenfeld and R. K. Jain, Direct Optical Probing of Integrated Circuits and High-Speed Devices
G. Plows, Electron-Beam Probing
A. M. Weiner and R. B. Marcus, Photoemissive Probing

Volume 29 Very High Speed Integrated Circuits: Gallium Arsenide LSI

M. Kuzuhara and T. Nazaki, Active Layer Formation by Ion Implantation
H. Hasimoto, Focused Ion Beam Implantation Technology
T. Nozaki and A. Higashisaka, Device Fabrication Process Technology
M. Ino and T. Takada, GaAs LSI Circuit Design
M. Hirayama, M. Ohmori, and K. Yamasaki, GaAs LSI Fabrication and Performance

Volume 30 Very High Speed Integrated Circuits: Heterostructure

H. Watanabe, T. Mizutani, and A. Usui, Fundamentals of Epitaxial Growth and Atomic Layer Epitaxy
S. Hiyamizu, Characteristics of Two-Dimensional Electron Gas in III–V Compound Heterostructures Grown by MBE
T. Nakanisi, Metalorganic Vapor Phase Epitaxy for High-Quality Active Layers
T. Nimura, High Electron Mobility Transistor and LSI Applications
T. Sugeta and T. Ishibashi, Hetero-Bipolar Transistor and LSI Application
H. Matsuedo, T. Tanaka, and M. Nakamura, Optoelectronic Integrated Circuits

Volume 31 Indium Phosphide: Crystal Growth and Characterization

J. P. Farges, Growth of Discoloration-Free InP
M. J. McCollum and G. E. Stillman, High Purity InP Grown by Hydride Vapor Phase Epitaxy
I. Inada and T. Fukuda, Direct Synthesis and Growth of Indium Phosphide by the Liquid Phosphorous Encapsulated Czochralski Method
O. Oda, K. Katagiri, K. Shinohara, S. Katsura, Y. Takahashi, K. Kainosho, K. Kohiro, and R. Hirano, InP Crystal Growth, Substrate Preparation and Evaluation
K. Tada, M. Tatsumi, M. Morioka, T. Araki, and T. Kawase, InP Substrates: Production and Quality Control
M. Razeghi, LP-MOCVD Growth, Characterization, and Application of InP Material
T. A. Kennedy and P. J. Lin-Chung, Stoichiometric Defects in InP

Volume 32   **Strained-Layer Superlattices: Physics**

T. P. Pearsall, Strained-Layer Superlattices
F. H. Pollack, Effects of Homogeneous Strain on the Electronic and Vibrational Levels in Semiconductors
J. Y. Marzin, J. M. Gerárd, P. Voisin, and J. A. Brum, Optical Studies of Strained III–V Heterolayers
R. People and S. A. Jackson, Structurally Induced States from Strain and Confinement
M. Jaros, Microscopic Phenomena in Ordered Superlattices

Volume 33   **Strained-Layer Superlattices: Material Science and Technology**

R. Hull and J. C. Bean, Principles and Concepts of Strained-Layer Epitaxy
W. J. Shaff, P. J. Tasker, M. C. Foisy, and L. F. Eastman, Device Applications of Strained-Layer Epitaxy
S. T. Picraux, B. L. Doyle, and J. Y. Tsao, Structure and Characterization of Strained-Layer Superlattices
E. Kasper and F. Schaffer, Group IV Compounds
D. L. Martin, Molecular Beam Epitaxy of IV–VI Compounds Heterojunction
R. L. Gunshor, L. A. Kolodziejski, A. V. Nurmikko, and N. Otsuka, Molecular Beam Epitaxy of I–VI Semiconductor Microstructures

Volume 34   **Hydrogen in Semiconductors**

J. I. Pankove and N. M. Johnson, Introduction to Hydrogen in Semiconductors
C. H. Seager, Hydrogenation Methods
J. I. Pankove, Hydrogenation of Defects in Crystalline Silicon
J. W. Corbett, P. Déak, U. V. Desnica, and S. J. Pearton, Hydrogen Passivation of Damage Centers in Semiconductors
S. J. Pearton, Neutralization of Deep Levels in Silicon
J. I. Pankove, Neutralization of Shallow Acceptors in Silicon
N. M. Johnson, Neutralization of Donor Dopants and Formation of Hydrogen-Induced Defects in n-Type Silicon
M. Stavola and S. J. Pearton, Vibrational Spectroscopy of Hydrogen-Related Defects in Silicon
A. D. Marwick, Hydrogen in Semiconductors: Ion Beam Techniques
C. Herring and N. M. Johnson, Hydrogen Migration and Solubility in Silicon
E. E. Haller, Hydrogen-Related Phenomena in Crystalline Germanium
J. Kakalios, Hydrogen Diffusion in Amorphous Silicon
J. Chevalier, B. Clerjaud, and B. Pajot, Neutralization of Defects and Dopants in III–V Semiconductors
G. G. DeLeo and W. B. Fowler, Computational Studies of Hydrogen-Containing Complexes in Semiconductors
R. F. Kiefl and T. L. Estle, Muonium in Semiconductors
C. G. Van de Walle, Theory of Isolated Interstitial Hydrogen and Muonium in Crystalline Semiconductors

Volume 35   **Nanostructured Systems**

M. Reed, Introduction
H. van Houten, C. W. J. Beenakker, and B. J. Wees, Quantum Point Contacts
G. Timp, When Does a Wire Become an Electron Waveguide?
M. Búttiker, The Quantum Hall Effects in Open Conductors
W. Hansen, J. P. Kotthaus, and U. Merkt, Electrons in Laterally Periodic Nanostructures

Volume 36 The Spectroscopy of Semiconductors

D. Heiman, Spectroscopy of Semiconductors at Low Temperatures and High Magnetic Fields
A. V. Nurmikko, Transient Spectroscopy by Ultrashort Laser Pulse Techniques
A. K. Ramdas and S. Rodriguez, Piezospectroscopy of Semiconductors
O. J. Glembocki and B. V. Shanabrook, Photoreflectance Spectroscopy of Microstructures
D. G. Seiler, C. L. Littler, and M. H. Wiler, One- and Two-Photon Magneto-Optical Spectroscopy of InSb and $Hg_{1-x}Cd_xTe$

Volume 37 The Mechanical Properties of Semiconductors

A.-B. Chen, A. Sher, and W. T. Yost, Elastic Constants and Related Properties of Semiconductor Compounds and Their Alloys
D. R. Clarke, Fracture of Silicon and Other Semiconductors
H. Siethoff, The Plasticity of Elemental and Compound Semiconductors
S. Guruswamy, K. T. Faber, and J. P. Hirth, Mechanical Behavior of Compound Semiconductors
S. Mahajan, Deformation Behavior of Compound Semiconductors
J. P. Hirth, Injection of Dislocations into Strained Multilayer Structures
D. Kendall, C. B. Fleddermann, and K. J. Malloy, Critical Technologies for the Micromatching of Silicon
J. Matsuba and K. Mokuya, Processing and Semiconductor Thermoelastic Behavior

Volume 38 Imperfections in III/V Materials

U. Scherz and M. Scheffler, Density-Functional Theory of sp-Bonded Defects in III/V Semiconductors
M. Kaminska and E. R. Weber, E12 Defect in GaAs
D. C. Look, Defects Relevant for Compensation in Semi-Insulating GaAs
R. C. Newman, Local Vibrational Mode Spectroscopy of Defects in III/V Compounds
A. M. Hennel, Transition Metals in III/V Compounds
K. J. Malloy and K. Khachaturyan, DX and Related Defects in Semiconductors
V. Swaminathan and A. S. Jordan, Dislocations in III/V Compounds
K. W. Nauka, Deep Level Defects in the Epitaxial III/V Materials

Volume 39 Minority Carriers in III–V Semiconductors: Physics and Applications

N. K. Dutta, Radiative Transition in GaAs and Other III–V Compounds
R. K. Ahrenkiel, Minority-Carrier Lifetime in III–V Semiconductors
T. Furuta, High Field Minority Electron Transport in p-GaAs
M. S. Lundstrom, Minority-Carrier Transport in III–V Semiconductors
R. A. Abram, Effects of Heavy Doping and High Excitation on the Band Structure of GaAs
D. Yevick and W. Bardyszewski, An Introduction to Non-Equilibrium Many-Body Analyses of Optical Processes in III–V Semiconductors

Volume 40 Epitaxial Microstructures

E. F. Schubert, Delta-Doping of Semiconductors: Electronic, Optical and Structural Properties of Materials and Devices

A. Gossard, M. Sundaram, and P. Hopkins, Wide Graded Potential Wells
P. Petroff, Direct Growth of Nanometer-Size Quantum Wire Superlattices
E. Kapon, Lateral Patterning of Quantum Well Heterostructures by Growth of Nonplanar Substrates
H. Temkin, D. Gershoni, and M. Panish, Optical Properties of $Ga_{1-x}In_xAs$/InP Quantum Wells

Volume 41 High Speed Heterostructure Devices

F. Capasso, F. Beltram, S. Sen, A. Pahlevi, and A. Y. Cho, Quantum Electron Devices: Physics and Applications
P. Solomon, D. J. Frank, S. L. Wright and F. Canora, GaAs-Gate Semiconductor-Insulator- Semiconductor FET
M. H. Hashemi and U. K. Mishra, Unipolar InP-Based Transistors
R. Kiehl, Complementary Heterostructure FET Integrated Circuits
T. Ishibashi, GaAs-Based and InP-Based Heterostructure Bipolar-Transistors
H. C. Liu and T. C. L. G. Sollner, High-Frequency-Tunneling Devices
H. Ohnishi, T. More, M. Takatsu, K. Imamura, and N. Yokoyama, Resonant-Tunneling Hot-Electron Transistors and Circuits

Volume 42 Oxygen in Silicon

F. Shimura, Introduction to Oxygen in Silicon
W. Lin, The Incorporation of Oxygen into Silicon Crystals
T. J. Schaffner and D. K. Schroder, Characterization Techniques for Oxygen in Silicon
W. M. Bullis, Oxygen Concentration Measurement
S. M. Hu, Intrinsic Point Defects in Silicon
B. Pajot, Some Atomic Configuration of Oxygen
J. Michel and L. C. Kimerling, Electrical Properties of Oxygen in Silicon
R. C. Newman and R. Jones, Diffusion of Oxygen in Silicon
T. Y. Tan and W. J. Taylor, Mechanisms of Oxygen Precipitation: Some Quantitative Aspects
M. Schrems, Simulation of Oxygen Precipitation
K. Simino and I. Yonenaga, Oxygen Effect on Mechanical Properties
W. Bergholz, Grown-in and Process-Induced Effects
F. Shimura, Intrinsic/Internal Gettering
H. Tsuya, Oxygen Effect on Electronic Device Performance

Volume 43 Semiconductors for Room Temperature Nuclear Detector Applications

R. B. James and T. E. Schlesinger, Introduction and Overview
L. S. Darken and C. E. Cox, High-Purity Germanium Detectors
A. Burger, D. Nason, L. Van den Berg, and M. Schieber, Growth of Mercuric Iodide
X. J. Bao, T. E. Schlesinger, and R. B. James, Electrical Properties of Mercuric Iodide
X. J. Bao, R. B. James, and T. E. Schlesinger, Optical Properties of Red Mercuric Iodide
M. Hage-Ali and P. Siffert, Growth Methods of CdTe Nuclear Detector Materials
M. Hage-Ali and P. Siffert, Characterization of CdTe Nuclear Detector Materials
M. Hage-Ali and P. Siffert, CdTe Nuclear Detectors and Applications
R. B. James, T. E. Schlesinger, J. Lund, and M. Schieber, $Cd_{1-x}Zn_x$Te Spectrometers for Gamma and X-Ray Applications

D. S. McGregor, J. E. Kammeraad, Gallium Arsenide Radiation Detectors and Spectrometers

J. C. Lund, F. Olschner, and A. Burger, Lead Iodide

M. R. Squillante and K. S. Shah, Other Materials: Status and Prospects

V. M. Gerrish, Characterization and Quantification of Detector Performance

J. S. Iwanczyk and B. E. Patt, Electronics for X-ray and Gamma Ray Spectrometers

M. Schieber, R. B. James and T. E. Schlesinger, Summary and Remaining Issues for Room Temperature Radiation Spectrometers

Volume 44 II–IV Blue/Green Light Emitters: Device Physics and Epitaxial Growth

J. Han and R. L. Gunshor, MBE Growth and Electrical Properties of Wide Bandgap ZnSe-based II–VI Semiconductors

S. Fujita and S. Fujita, Growth and Characterization of ZnSe-based II–VI Semiconductors by MOVPE

E. Ho and L. A. Kolodziejski, Gaseous Source UHV Epitaxy Technologies for Wide Bandgap II–VI Semiconductors

C. G. Van de Walle, Doping of Wide-Band-Gap II–VI Compounds – Theory

R. Cingolani, Optical Properties of Excitons in ZnSe-Based Quantum Well Heterostructures

A. Ishibashi and A. V. Nurmikko, II–VI Diode Lasers: A Current View of Device Performance and Issues

S. Guha and J. Petruzello, Defects and Degradation in Wide-Gap II–VI-based Structure and Light Emitting Devices

Volume 45 Effect of Disorder and Defects in Ion-Implanted Semiconductors: Electrical and Physiochemical Characterization

H. Ryssel, Ion Implantation into Semiconductors: Historical Perspectives

You-Nian Wang and Teng-Cai Ma, Electronic Stopping Power for Energetic Ions in Solids

S. T. Nakagawa, Solid Effect on the Electronic Stopping of Crystalline Target and Application to Range Estimation

G. Müller, S. Kalbitzer, and G. N. Greaves, Ion Beams in Amorphous Semiconductor Research

J. Boussey-Said, Sheet and Spreading Resistance Analysis of Ion Implanted and Annealed Semiconductors

M. L. Polignano and G. Queirolo, Studies of the Stripping Hall Effect in Ion-Implanted Silicon

J. Sroemenos, Transmission Electron Microscopy Analyses

R. Nipoti and M. Servidori, Rutherford Backscattering Studies of Ion Implanted Semiconductors

P. Zaumseil, X-ray Diffraction Techniques

Volume 46 Effect of Disorder and Defects in Ion-Implanted Semiconductors: Optical and Photothermal Characterization

M. Fried, T. Lohner, and J. Gyulai, Ellipsometric Analysis

A. Seas and C. Christofides, Transmission and Reflection Spectroscopy on Ion Implanted Semiconductors

A. Othonos and C. Christofides, Photoluminescence and Raman Scattering of Ion Implanted Semiconductors. Influence of Annealing

C. Christofides, Photomodulated Thermoreflectance Investigation of Implanted Wafers. Annealing Kinetics of Defects

U. Zammit, Photothermal Deflection Spectroscopy Characterization of Ion-Implanted and Annealed Silicon Films

A. Mandelis, A. Budiman, and M. Vargas, Photothermal Deep-Level Transient Spectroscopy of Impurities and Defects in Semiconductors

R. Kalish and S. Charbonneau, Ion Implantation into Quantum-Well Structures

A. M. Myasnikov and N. N. Gerasimenko, Ion Implantation and Thermal Annealing of III–V Compound Semiconducting Systems: Some Problems of III–V Narrow Gap Semiconductors

Volume 47 Uncooled Infrared Imaging Arrays and Systems

R. G. Buser and M. P. Tompsett, Historical Overview

P. W. Kruse, Principles of Uncooled Infrared Focal Plane Arrays

R. A. Wood, Monolithic Silicon Microbolometer Arrays

C. M. Hanson, Hybrid Pyroelectric-Ferroelectric Bolometer Arrays

D. L. Polla and J. R. Choi, Monolithic Pyroelectric Bolometer Arrays

N. Teranishi, Thermoelectric Uncooled Infrared Focal Plane Arrays

M. F. Tompsett, Pyroelectric Vidicon

T. W. Kenny, Tunneling Infrared Sensors

J. R. Vig, R. L Filler, and Y. Kim, Application of Quartz Microresonators to Uncooled Infrared Imaging Arrays

P. W. Kruse, Application of Uncooled Monolithic Thermoelectric Linear Arrays to Imaging Radiometers

Volume 48 High Brightness Light Emitting Diodes

G. B. Stringfellow, Materials Issues in High-Brightness Light-Emitting Diodes

M.G. Craford, Overview of Device Issues in High-Brightness Light-Emitting Diodes

F. M. Steranka, AlGaAs Red Light Emitting Diodes

C. H. Chen, S. A. Stockman, M. J. Peanasky, and C. P. Kuo, OMVPE Growth of AlGaInP for High Efficiency Visible Light-Emitting Diodes

F. A. Kish and R. M. Fletcher, AlGaInP Light-Emitting Diodes

M. W. Hodapp, Applications for High Brightness Light-Emitting Diodes

I. Akasaki and H. Amano, Organometallic Vapor Epitaxy of GaN for High Brightness Blue Light Emitting Diodes

S. Nakamura, Group III–V Nitride Based Ultraviolet-Blue-Green-Yellow Light-Emitting Diodes and Laser Diodes

Volume 49 Light Emission in Silicon: from Physics to Devices

D. J. Lockwood, Light Emission in Silicon

G. Abstreiter, Band Gaps and Light Emission in Si/SiGe Atomic Layer Structures

T. G. Brown and D. G. Hall, Radiative Isoelectronic Impurities in Silicon and Silicon-Germanium Alloys and Superlattices

J. Michel, L. V. C. Assali, M. T. Morse, and L. C. Kimerling, Erbium in Silicon

Y. Kanemitsu, Silicon and Germanium Nanoparticles

P. M. Fauchet, Porous Silicon: Photoluminescence and Electroluminescent Devices

C. Delerue, G. Allan, and M. Lannoo, Theory of Radiative and Nonradiative Processes in Silicon Nanocrystallites

L. Brus, Silicon Polymers and Nanocrystals

Volume 50   Gallium Nitride (GaN)

J. I. Pankove and T. D. Moustakas, Introduction
S. P. DenBaars and S. Keller, Metalorganic Chemical Vapor Deposition (MOCVD) of Group III Nitrides
W. A. Bryden and T. J. Kistenmacher, Growth of Group III–A Nitrides by Reactive Sputtering
N. Newman, Thermochemistry of III–N Semiconductors
S. J. Pearton and R. J. Shul, Etching of III Nitrides
S. M. Bedair, Indium-based Nitride Compounds
A. Trampert, O. Brandt, and K. H. Ploog, Crystal Structure of Group III Nitrides
H. Morkoç, F. Hamdani, and A. Salvador, Electronic and Optical Properties of III–V Nitride based Quantum Wells and Superlattices
K. Doverspike and J. I. Pankove, Doping in the III-Nitrides
T. Suski and P. Perlin, High Pressure Studies of Defects and Impurities in Gallium Nitride
B. Monemar, Optical Properties of GaN
W. R. L. Lambrecht, Band Structure of the Group III Nitrides
N. E. Christensen and P. Perlin, Phonons and Phase Transitions in GaN
S. Nakamura, Applications of LEDs and LDs
I. Akasaki and H. Amano, Lasers
J. A. Cooper, Jr., Nonvolatile Random Access Memories in Wide Bandgap Semiconductors

Volume 51A   Identification of Defects in Semiconductors

G. D. Watkins, EPR and ENDOR Studies of Defects in Semiconductors
J.-M. Spaeth, Magneto-Optical and Electrical Detection of Paramagnetic Resonance in Semiconductors
T. A. Kennedy and E. R. Claser, Magnetic Resonance of Epitaxial Layers Detected by Photoluminescence
K. H. Chow, B. Hitti, and R. F. Kiefl, µSR on Muonium in Semiconductors and Its Relation to Hydrogen
K. Saarinen, P. Hautojärvi, and C. Corbel, Positron Annihilation Spectroscopy of Defects in Semiconductors
R. Jones and P. R. Briddon, The *Ab Initio* Cluster Method and the Dynamics of Defects in Semiconductors

Volume 51B   Identification Defects in Semiconductors

G. Davies, Optical Measurements of Point Defects
P. M. Mooney, Defect Identification Using Capacitance Spectroscopy
M. Stavola, Vibrational Spectroscopy of Light Element Impurities in Semiconductors
P. Schwander, W. D. Rau, C. Kisielowski, M. Gribelyuk, and A. Ourmazd, Defect Processes in Semiconductors Studied at the Atomic Level by Transmission Electron Microscopy
N. D. Jager and E. R. Weber, Scanning Tunneling Microscopy of Defects in Semiconductors

Volume 52   SiC Materials and Devices

K. Järrendahl and R. F. Davis, Materials Properties and Characterization of SiC
V. A. Dmitriev and M. G. Spencer, SiC Fabrication Technology: Growth and Doping
V. Saxena and A. J. Steckl, Building Blocks for SiC Devices: Ohmic Contacts, Schottky Contacts, and p-n Junctions
M. S. Shur, SiC Transistors
C. D. Brandt, R. C. Clarke, R. R. Siergiej, J. B. Casady, A. W. Morse, S. Sriram, and A. K. Agarwal, SiC for Applications in High-Power Electronics

R. J. Trew, SiC Microwave Devices

J. Edmond, H. Kong, G. Negley, M. Leonard, K. Doverspike, W. Weeks, A. Suvorov, D. Waltz, and C. Carter, Jr., SiC-Based UV Photodiodes and Light-Emitting Diodes

H. Morkoç, Beyond Silicon Carbide! III–V Nitride-Based Heterostructures and Devices

Volume 53 Cumulative Subjects and Author Index Including Tables of Contents for Volumes 1–50

Volume 54 High Pressure in Semiconductor Physics I

W. Paul, High Pressure in Semiconductor Physics: A Historical Overview

N. E. Christensen, Electronic Structure Calculations for Semiconductors Under Pressure

R. J. Neimes and M. I. McMahon, Structural Transitions in the Group IV, III–V and II–VI Semiconductors Under Pressure

A. R. Goni and K. Syassen, Optical Properties of Semiconductors Under Pressure

P. Trautman, M. Baj, and J. M. Baranowski, Hydrostatic Pressure and Uniaxial Stress in Investigations of the EL2 Defect in GaAs

M. Li and P. Y. Yu, High-Pressure Study of DX Centers Using Capacitance Techniques

T. Suski, Spatial Correlations of Impurity Charges in Doped Semiconductors

N. Kuroda, Pressure Effects on the Electronic Properties of Diluted Magnetic Semiconductors

Volume 55 High Pressure in Semiconductor Physics II

D. K. Maude and J. C. Portal, Parallel Transport in Low-Dimensional Semiconductor Structures

P. C. Klipstein, Tunneling Under Pressure: High-Pressure Studies of Vertical Transport in Semiconductor Heterostructures

E. Anastassakis and M. Cardona, Phonons, Strains, and Pressure in Semiconductors

F. H. Pollak, Effects of External Uniaxial Stress on the Optical Properties of Semiconductors and Semiconductor Microstructures

A. R. Adams, M. Silver, and J. Allam, Semiconductor Optoelectronic Devices

S. Porowski and I. Grzegory, The Application of High Nitrogen Pressure in the Physics and Technology of III–N Compounds

M. Yousuf, Diamond Anvil Cells in High Pressure Studies of Semiconductors

Volume 56 Germanium Silicon: Physics and Materials

J. C. Bean, Growth Techniques and Procedures

D. E. Savage, F. Liu, V. Zielasek, and M. G. Lagally, Fundamental Crystal Growth Mechanisms

R. Hull, Misfit Strain Accommodation in SiGe Heterostructures

M. J. Shaw and M. Jaros, Fundamental Physics of Strained Layer GeSi: Quo Vadis?

F. Cerdeira, Optical Properties

S. A. Ringel and P. N. Grillot, Electronic Properties and Deep Levels in Germanium-Silicon

J. C. Campbell, Optoelectronics in Silicon and Germanium Silicon

K. Eberl, K. Brunner, and O. G. Schmidt, $Si_{1-y}C_y$ and $Si_{1-x-y}Ge_xC_y$ Alloy Layers

Volume 57   Gallium Nitride (GaN) II

R. J. Molnar, Hydride Vapor Phase Epitaxial Growth of III–V Nitrides

T. D. Moustakas, Growth of III–V Nitrides by Molecular Beam Epitaxy

Z. Liliental-Weber, Defects in Bulk GaN and Homoepitaxial Layers

C. G. Van de Walk and N. M. Johnson, Hydrogen in III–V Nitrides

W. Götz and N. M. Johnson, Characterization of Dopants and Deep Level Defects in Gallium Nitride

B. Gil, Stress Effects on Optical Properties

C. Kisielowski, Strain in GaN Thin Films and Heterostructures

J. A. Miragliotta and D. K. Wickenden, Nonlinear Optical Properties of Gallium Nitride

B. K. Meyer, Magnetic Resonance Investigations on Group III–Nitrides

M. S. Shur and M. Asif Khan, GaN and AlGaN Ultraviolet Detectors

C. H. Qiu, J. I. Pankove, and C. Rossington, II–V Nitride-Based X-ray Detectors

Volume 58   Nonlinear Optics in Semiconductors I

A. Kost, Resonant Optical Nonlinearities in Semiconductors

E. Garmire, Optical Nonlinearities in Semiconductors Enhanced by Carrier Transport

D. S. Chemla, Ultrafast Transient Nonlinear Optical Processes in Semiconductors

M. Sheik-Bahae and E. W. Van Stryland, Optical Nonlinearities in the Transparency Region of Bulk Semiconductors

J. E. Millerd, M. Ziari, and A. Partovi, Photorefractivity in Semiconductors

Volume 59   Nonlinear Optics in Semiconductors II

J. B. Khurgin, Second Order Nonlinearities and Optical Rectification

K. L. Hall, E. R. Thoen, and E. P. Ippen, Nonlinearities in Active Media

E. Hanamura, Optical Responses of Quantum Wires/Dots and Microcavities

U. Keller, Semiconductor Nonlinearities for Solid-State Laser Modelocking and Q-Switching

A. Miller, Transient Grating Studies of Carrier Diffusion and Mobility in Semiconductors

Volume 60   Self-Assembled InGaAs/GaAs Quantum Dots

Mitsuru Sugawara, Theoretical Bases of the Optical Properties of Semiconductor Quantum Nano-Structures

Yoshiaki Nakata, Yoshihiro Sugiyama, and Mitsuru Sugawara, Molecular Beam Epitaxial Growth of Self-Assembled InAs/GaAs Quantum Dots

Kohki Mukai, Mitsuru Sugawara, Mitsuru Egawa, and Nobuyuki Ohtsuka, Metalorganic Vapor Phase Epitaxial Growth of Self-Assembled InGaAs/GaAs Quantum Dots Emitting at 1.3 μm

Kohki Mukai and Mitsuru Sugawara, Optical Characterization of Quantum Dots

Kohki Mukai and Mitsuru Sugawara, The Photon Bottleneck Effect in Quantum Dots

Hajime Shoji, Self-Assembled Quantum Dot Lasers

Hiroshi Ishikawa, Applications of Quantum Dot to Optical Devices

Mitsuru Sugawara, Kohki Mukai, Hiroshi Ishikawa, Koji Otsubo, and Yoshiaki Nakata, The Latest News

Volume 61   Hydrogen in Semiconductors II

Norbert H. Nickel, Introduction to Hydrogen in Semiconductors II

Noble M. Johnson and Chris G. Van de Walle, Isolated Monatomic Hydrogen in Silicon

Yurij V. Gorelkinskii, Electron Paramagnetic Resonance Studies of Hydrogen and Hydrogen-Related Defects in Crystalline Silicon

Norbert H. Nickel, Hydrogen in Polycrystalline Silicon

Wolfhard Beyer, Hydrogen Phenomena in Hydrogenated Amorphous Silicon

Chris G. Van de Walle, Hydrogen Interactions with Polycrystalline and Amorphous Silicon–Theory

Karen M. McManus Rutledge, Hydrogen in Polycrystalline CVD Diamond

Roger L. Lichti, Dynamics of Muonium Diffusion, Site Changes and Charge-State Transitions

Matthew D. McCluskey and Eugene E. Haller, Hydrogen in III–V and II–VI Semiconductors

S. J. Pearton and J. W. Lee, The Properties of Hydrogen in GaN and Related Alloys

Jörg Neugebauer and Chris G. Van de Walle, Theory of Hydrogen in GaN

Volume 62 Intersubband Transitions in Quantum Wells: Physics and Device Applications I

Manfred Helm, The Basic Physics of Intersubband Transitions

Jerome Faist, Carlo Sirtori, Federico Capasso, Loren N. Pfeiffer, Ken W. West, Deborah L. Sivco, and Alfred Y. Cho, Quantum Interference Effects in Intersubband Transitions

H. C. Liu, Quantum Well Infrared Photodetector Physics and Novel Devices

S. D. Gunapala and S. V. Bandara, Quantum Well Infrared Photodetector (QWIP) Focal Plane Arrays

Volume 63 Chemical Mechanical Polishing in Si Processing

Frank B. Kaufman, Introduction

Thomas Bibby and Karey Holland, Equipment

John P. Bare, Facilitization

Duane S. Boning and Okumu Ouma, Modeling and Simulation

Shin Hwa Li, Bruce Tredinnick, and Mel Hoffman, Consumables I: Slurry

Lee M. Cook, CMP Consumables II: Pad

François Tardif, Post-CMP Clean

Shin Hwa Li, Tara Chhatpar, and Frederic Robert, CMP Metrology

Shin Hwa Li, Visun Bucha, and Kyle Wooldridge, Applications and CMP-Related Process Problems

Volume 64 Electroluminescence I

M. G. Craford, S. A. Stockman, M. J. Peansky, and F. A. Kish, Visible Light-Emitting Diodes

H. Chui, N. F. Gardner, P. N. Grillot, J. W. Huang, M. R. Krames, and S. A. Maranowski, High-Efficiency AIGaInP Light-Emitting Diodes

R. S. Kern, W. Götz, C. H. Chen, H. Liu, R. M. Fletcher, and C. P. Kuo, High-Brightness Nitride-Based Visible-Light-Emitting Diodes

Yoshiharu Sato, Organic LED System Considerations

V. Bulović, P. E. Burrows, and S. R. Forrest, Molecular Organic Light-Emitting Devices

Volume 65 Electroluminescence II

V. Bulović and S. R. Forrest, Polymeric and Molecular Organic Light Emitting Devices: A Comparison

Regina Mueller-Mach and Gerd O. Mueller, Thin Film Electroluminescence
Markku Leskelä, Wei-Min Li, and Mikko Ritala, Materials in Thin Film Electroluminescent Devices
Kristiaan Neyts, Microcavities for Electroluminescent Devices

Volume 66 Intersubband Transitions in Quantum Wells: Physics and Device Applications II

Jerome Faist, Federico Capasso, Carlo Sirtori, Deborah L. Sivco, and Alfred Y. Cho, Quantum Cascade Lasers
Federico Capasso, Carlo Sirtori, D. L. Sivco, and A. Y. Cho, Nonlinear Optics in Coupled-Quantum-Well Quasi-Molecules
Karl Unterrainer, Photon-Assisted Tunneling in Semiconductor Quantum Structures
P. Haring Bolivar, T. Dekorsy, and H. Kurz, Optically Excited Bloch Oscillations–Fundamentals and Application Perspectives

Volume 67 Ultrafast Physical Processes in Semiconductors

Alfred Leitenstorfer and Alfred Laubereau, Ultrafast Electron-Phonon Interactions in Semiconductors: Quantum Kinetic Memory Effects
Christoph Lienau and Thomas Elsaesser, Spatially and Temporally Resolved Near-Field Scanning Optical Microscopy Studies of Semiconductor Quantum Wires
K. T. Tsen, Ultrafast Dynamics in Wide Bandgap Wurtzite GaN
J. Paul Callan, Albert M.-T. Kim, Christopher A. D. Roeser, and Eriz Mazur, Ultrafast Dynamics and Phase Changes in Highly Excited GaAs
Hartmut Hang, Quantum Kinetics for Femtosecond Spectroscopy in Semiconductors
T. Meier and S. W. Koch, Coulomb Correlation Signatures in the Excitonic Optical Nonlinearities of Semiconductors
Roland E. Allen, Traian Dumitrică, and Ben Torralva, Electronic and Structural Response of Materials to Fast, Intense Laser Pulses
E. Gornik and R. Kersting, Coherent THz Emission in Semiconductors

Volume 68 Isotope Effects in Solid State Physics

Vladimir G. Plekhanov, Elastic Properties; Thermal Properties; Vibrational Properties; Raman Spectra of Isotopically Mixed Crystals; Excitons in LiH Crystals; Exciton–Phonon Interaction; Isotopic Effect in the Emission Spectrum of Polaritons; Isotopic Disordering of Crystal Lattices; Future Developments and Applications; Conclusions

Volume 69 Recent Trends in Thermoelectric Materials Research I

H. Julian Goldsmid, Introduction
Terry M. Tritt and Valerie M. Browning, Overview of Measurement and Characterization Techniques for Thermoelectric Materials
Mercouri G. Kanatzidis, The Role of Solid-State Chemistry in the Discovery of New Thermoelectric Materials
B. Lenoir, H. Scherrer, and T. Caillat, An Overview of Recent Developments for BiSb Alloys
Citrad Uher, Skutterudities: Prospective Novel Thermoelectrics
George S. Nolas, Glen A. Slack, and Sandra B. Schujman, Semiconductor Clathrates: A Phonon Glass Electron Crystal Material with Potential for Thermoelectric Applications

Volume 70 Recent Trends in Thermoelectric Materials Research II

Brian C. Sales, David G. Mandrus, and Bryan C. Chakoumakos, Use of Atomic Displacement Parameters in Thermoelectric Materials Research

S. Joseph Poon, Electronic and Thermoelectric Properties of Half-Heusler Alloys

Terry M. Tritt, A. L. Pope, and J. W. Kolis, Overview of the Thermoelectric Properties of Quasicrystalline Materials and Their Potential for Thermoelectric Applications

Alexander C. Ehrlich and Stuart A. Wolf, Military Applications of Enhanced Thermoelectrics

David J. Singh, Theoretical and Computational Approaches for Identifying and Optimizing Novel Thermoelectric Materials

Terry M. Tritt and R. T. Littleton, IV, Thermoelectric Properties of the Transition Metal Pentatellurides: Potential Low-Temperature Thermoelectric Materials

Franz Freibert, Timothy W. Darling, Albert Migliori, and Stuart A. Trugman, Thermomagnetic Effects and Measurements

M. Bartkowiak and G. D. Mahan, Heat and Electricity Transport Through Interfaces

Volume 71 Recent Trends in Thermoelectric Materials Research III

M. S. Dresselhaus, Y.-M. Lin, T. Koga, S. B. Cronin, O. Rabin, M. R. Black, and G. Dresselhaus, Quantum Wells and Quantum Wires for Potential Thermoelectric Applications

D. A. Broido and T. L. Reinecke, Thermoelectric Transport in Quantum Well and Quantum Wire Superlattices

G. D. Mahan, Thermionic Refrigeration

Rama Venkatasubramanian, Phonon Blocking Electron Transmitting Superlattice Structures as Advanced Thin Film Thermoelectric Materials

G. Chen, Phonon Transport in Low-Dimensional Structures

Volume 72 Silicon Epitaxy

S. Acerboni, ST Microelectronics, CFM-AGI Department, Agrate Brianza, Italy

V.-M. Airaksinen, Okmetic Oyj R&D Department, Vantaa, Finland

G. Beretta, ST Microelectronics, DSG Epitaxy Catania Department, Catania, Italy

C. Cavallotti, Dipartimento di Chimica Fisica Applicata, Politecnico di Milano, Milano, Italy

D. Crippa, MEMC Electronic Materials, Epitaxial and CVD Department, Operations Technology Division, Novara, Italy

D. Dutartre, ST Microelectronics, Central R&D, Crolles, France

Srikanth Kommu, MEMC Electronic Materials inc., EPI Technology Group, St. Peters, Missouri

M. Masi, Dipartimento di Chimica Fisica Applicata, Politecnico di Milano, Milano, Italy

D. J. Meyer, ASM Epitaxy, Phoenix, Arizona

J. Murota, Research Institute of Electrical Communication, Laboratory for Electronic Intelligent Systems, Tohoku University, Sendai, Japan

V. Pozzetti, LPE Epitaxial Technologies, Bollate, Italy

A. M. Rinaldi, MEMC Electronic Materials, Epitaxial and CVD Department, Operations Technology Division, Novara, Italy

Y. Shiraki, Research Center for Advanced Science and Technology (RCAST), University of Tokyo, Tokyo, Japan

Volume 73 Processing and Properties of Compound Semiconductors

S. J. Pearton, Introduction

Eric Donkor, Gallium Arsenide Heterostructures
Annamraju Kasi Viswanatli, Growth and Optical Properties of GaN
D. Y. C. Lie and K. L. Wang, SiGe/Si Processing
S. Kim and M. Razeghi, Advances in Quantum Dot Structures
Walter P. Gomes, Wet Etching of III–V Semiconductors

Volume 74 Silicon-Germanium Strained Layers and Heterostructures

S. C. Jain and M. Willander, Introduction; Strain, Stability, Reliability and Growth; Mechanism of Strain Relaxation; Strain, Growth, and TED in SiGeC Layers; Bandstructure and Related Properties; Heterostructure Bipolar Transistors; FETs and Other Devices

Volume 75 Laser Crystallization of Silicon

Norbert H. Nickel, Introduction to Laser Crystallization of Silicon
Costas P. Grigoropoidos, Seung-Jae Moon and Ming-Hong Lee, Heat Transfer and Phase Transformations in Laser Melting and Recrystallization of Amorphous Thin Si Films
Robert Černý and Petr Přikryl, Modeling Laser-Induced Phase-Change Processes: Theory and Computation
Paulo V. Santos, Laser Interference Crystallization of Amorphous Films
Philipp Lengsfeld and Norbert H. Nickel, Structural and Electronic Properties of Laser-Crystallized Poly-Si

Volume 76 Thin-Film Diamond I

X. Jiang, Textured and Heteroepitaxial CVD Diamond Films
Eberhard Blank, Structural Imperfections in CVD Diamond Films
R. Kalish, Doping Diamond by Ion-Implantation
A. Deneuville, Boron Doping of Diamond Films from the Gas Phase
S. Koizumi, n-Type Diamond Growth
C. E. Nebel, Transport and Defect Properties of Intrinsic and Boron-Doped Diamond
Miloš Nesládek, Ken Haenen and Milan Vaněček, Optical Properties of CVD Diamond
Rolf Sauer, Luminescence from Optical Defects and Impurities in CVD Diamond

Volume 77 Thin-Film Diamond II

Jacques Chevallier, Hydrogen Diffusion and Acceptor Passivation in Diamond
Jürgen Ristein, Structural and Electronic Properties of Diamond Surfaces
John C. Angus, Yuri V. Pleskov and Sally C. Eaton, Electrochemistry of Diamond
Greg M. Swain, Electroanalytical Applications of Diamond Electrodes
Werner Haenni, Philippe Rychen, Matthyas Fryda and Christos Comninellis, Industrial Applications of Diamond Electrodes
Philippe Bergonzo and Richard B. Jackman, Diamond-Based Radiation and Photon Detectors
Hiroshi Kawarada, Diamond Field Effect Transistors Using H-Terminated Surfaces
Shinichi Shikata and Hideaki Nakahata, Diamond Surface Acoustic Wave Device

Volume 78   **Semiconducting Chalcogenide Glass I**

V. S. Minaev and S. P. Timoshenkov, Glass-Formation in Chalcogenide Systems and Periodic System

A. Popov, Atomic Structure and Structural Modification of Glass

V. A. Funtikov, Eutectoidal Concept of Glass Structure and Its Application in Chalcogenide Semiconductor Glasses

V. S. Minaev, Concept of Polymeric Polymorphous-Crystalloid Structure of Glass and Chalcogenide Systems: Structure and Relaxation of Liquid and Glass

Volume 79   **Semiconducting Chalcogenide Glass II**

M. D. Bal'makov, Information Capacity of Condensed Systems

A. Česnys, G. Juška and E. Montrimas, Charge Carrier Transfer at High Electric Fields in Noncrystalline Semiconductors

Andrey S. Glebov, The Nature of the Current Instability in Chalcogenide Vitreous Semiconductors

A. M. Andriesh, M. S. Iovu and S. D. Shutov, Optical and Photoelectrical Properties of Chalcogenide Glasses

V. Val. Sobolev and V. V. Sobolev, Optical Spectra of Arsenic Chalcogenides in a Wide Energy Range of Fundamental Absorption

Yu. S. Tver'yanovich, Magnetic Properties of Chalcogenide Glasses

Volume 80   **Semiconducting Chalcogenide Glass III**

Andrey S. Glebov, Electronic Devices and Systems Based on Current Instability in Chalcogenide Semiconductors

Dumitru Tsiulyanu, Heterostructures on Chalcogenide Glass and Their Applications

E. Bychkov, Yu. Tveryanovich and Yu. Vlasov, Ion Conductivity and Sensors

Yu. S. Tver'yanovich and A. Tverjanovich, Rare-earth Doped Chalcogenide Glass

M. F. Churbanov and V. G. Plotnichenko, Optical Fibers from High-purity Arsenic Chalcogenide Glasses